Dendroecology

PRINCIPLES AND PRACTICE

Richard L. Phipps & Thomas M. Yanosky

J.ROSS PUBLISHING

For Library of Congress Cataloging-in-Publication Data, please see the WAV section of the publisher's website at www.jrosspub.com/wav.

This book is dedicated to my twin sons, Ed and Joe, and to my daughter, Shelley, as she begins her career as a chemical engineer.

—Tom Yanosky

CONTENTS

PREFACE

When introducing a unit on botany, we routinely taught our students that the structure and function of plants can best be understood if they kept two things in mind—that plants make their own food and that they are stuck in one place. Unlike animals that can avoid harsh environmental conditions by seeking shelter or simply moving to a more hospitable location, trees have instead evolved structurally and physiologically to survive where they germinate. At any site where a tree grows, temperature and water availability (not to mention the frequency of damaging winds, ice storms, and floods) vary not only within any given year but are likely to be extreme when considered on time scales of decades or centuries. Were this not true, we would expect all trees to show virtually no yearly ring-width variability during the course of their lifetimes.

Even within a given locale, slight differences in site factors can result in noticeable differences in the form and species composition of forests. Trees of the same species in close proximity, and thus subject to the same climatic conditions, may nevertheless differ greatly in size and radial growth simply because one group grows on deep, rich soil whereas the other is rooted in thin, rocky soil. The early builders of wooden structures in Europe and North America selected tall, straight-boled trees from better soils and neglected trees of poorer form growing on rocky slopes. This was a matter of practicality. Some of the early settlers surely took notice of differences in nature and, based on their world, tried to explain them. Our understanding of limiting factors has progressed considerably since then.

Yet, it is the limitation of environmental factors controlling tree growth that forms the basis for dendrochronological investigations. A number of studies collected wood samples from old trees with considerable year-to-year variations in ring width. After removing the typical trend of decreasing ring widths as trees age, the resulting *climatic component* of the ring series was compared to recent climatic data (typically precipitation or temperature) and used to reconstruct the variable in question; that is, to estimate the climatic variable before the period of instrumented records, often over hundreds of years. Applications of this kind, however, are only one use of tree growth to make environmental inferences. For example, a forester or environmental engineer may be less interested in past climate but rather may wish to know if trees can be used to infer the impact of recent land use, tree pathology, or the contamination of soils or ground water. In other words, evidence for any environmental factor affecting the growth of a tree is often preserved in its subsequent radial growth and thus affords a historical perspective unavailable by other means. Going forward in time, planners may desire to use the future growth of trees as *biomonitors* to evaluate the progress of reclamation efforts.

In this book, we have attempted to show how the age and radial growth of trees under a variety of different conditions can be used to investigate basic questions about the ecological environment. Although it was not our intent to write a strictly how-to text or to summarize the vast literature on this subject, we have tried to give specific examples, many from our own work, that hopefully will offer points of departure for others wishing to use this methodology to answer their own research questions. We cheerfully concede that we have only scratched the surface with regard to the range of applications to ecological questions and, accordingly, we have mentioned numerous areas that we believe merit further study. We hope that this volume will stimulate interest in the field and be of practical use to the scientific community as well.

ACKNOWLEDGMENTS

We sincerely appreciate all the assistance and advice that was received in the preparation of this book. During more than four decades, there have been many coworkers and assistants associated with the research lab of the botanical studies group of the National Research Program of the Water Resources Division of the United States Geological Survey (USGS). From this group we should particularly like to acknowledge Robert S. Sigafoos, Cliff R. Hupp, Esther Flint, and John Whiton. Special thanks are due to Don Vroblesky and Mike Field also with the USGS and Malcolm Cleaveland and David Stahle of the University of Arkansas. We would like to acknowledge students at George Washington University, George Mason University, and the University of Illinois for their helpful reactions to portions of early drafts. Tom Yanosky would like to acknowledge and thank Drs. Harlan Banks and Lawrence Hamilton of Cornell University; Drs. Larry Rockwood, Melissa Stanley, and Judith Skog of George Mason University; and the esteemed plant physiologist Dr. Robert Weintraub of George Washington University with whom I had so many enjoyable Friday afternoons discussing botany and life. Finally, we both wish to acknowledge the support and guidance afforded to us by Gwen Eyeington of J. Ross Publishing.

ABOUT THE AUTHORS

Richard L. Phipps received his BS in Botany from Eastern Illinois University, an MS in Botany from The Ohio State University (OSU), and his PhD in botany with a specialization in ecology from OSU. He worked as a research botanist for the United States Geological Survey (1960–1991), establishing and operating a tree-ring laboratory as a project chief in the National Research Program of the Water Resources Division. Additionally, he taught special courses in dendrochronology and forest ecology at George Washington University (GWU), George Mason University (GMU) (adjunct professor), and the University of Illinois (visiting professor). He was a major research advisor for four PhD candidates (three while at GWU, one while at GMU) and numerous MS candidates while at GMU. His research interests included tree growth, tree physiology and development, and general forest ecology. Outside interests have included beekeeping, apitherapy, and small engine mechanics. He now divides his time between Fairfax, Virginia, and Nags Head, North Carolina.

Thomas M. Yanosky received a BS in Biology from Cornell University, an MS in Biology from George Mason University, and a PhD in Plant Physiology and Anatomy from George Washington University. For 32 years Tom worked as a research botanist at the United States Geological Survey where he conducted studies at the interface of plant biology and surface- and ground-water hydrology. He published the first peer-reviewed study using digital image analysis to identify and measure anatomical features within tree rings. In addition, he conducted some of the initial studies using the multi-element concentrations within tree rings to estimate the spatial extent and historical onset of contaminated ground water and soils. As a Lecturer at George Mason University for 17 years, he taught undergraduate and graduate courses in the Department of Biology. Tom was also an associate editor for the International Association of Wood Anatomists Bulletin. He is a member of American Mensa and has life-long interests in lepidoptery, art, and 19th century English and American poetry. He resides in Canyon Lake, Texas, with his wife, Rhonda.

1

RING FORMATION

While helping cut trees on our family farm when I (RLP) was a kid, my father showed me tree rings on freshly cut tree stumps. He pointed out that because a new ring was added each year, the age of a tree could be determined by simply counting the rings. In addition, he noted that a wide ring meant a good year and a narrow ring meant a poor year for tree growth. I doubt that he knew this because he had read or heard about some new scientific work being done in Arizona—I think he was telling me something that was generally known for perhaps hundreds or thousands of years.

The intent of this chapter is to describe tree rings. First, we will describe the basic types of rings. We will then present some of the basic concepts of tree rings in three dimensions. Finally, we will discuss the timing of ring formation. Of note, for North American trees, throughout this volume, we have used the nomenclature of American botanist Elbert Little.

ARIZONA AND A. E. DOUGLASS

Arizona figures prominently in the history of tree-ring research in the United States because of Andrew Ellicott (A. E.) Douglass, who founded the Laboratory of Tree-Ring Research (LTRR) at the University of Arizona. There are probably numerous variations of the story as to how Douglass became interested in tree rings. In the early 1900s, Douglass, an astronomer from Harvard, was working at the observatory at Flagstaff. According to one variation of the story, Douglass was fascinated by tree rings that he noticed on freshly cut stumps in a ponderosa pine tree stand. In counting the rings on a stump, he was struck by the observation that the ring-width pattern had a quasi-resemblance to another pattern with which he was familiar; that is, the pattern of known variation in sun spot activity. One can imagine that he thought this to be an amusing coincidence. When he looked at another stump, he found that its pattern of rings was similar enough to the first that the rings of extreme size seemed to have formed in the same years. He subsequently noticed that the ring patterns on all of the stumps had at least some pattern characteristics in common. Furthermore, when he examined the stumps in another recently cut stand, he found similarities in patterns of wide and narrow rings with the first stand.

Sometime later, while looking at tree rings on a structural beam in an Indian ruin, Douglass wondered if it might be possible to match the pattern on the beam with patterns from very old living trees, thereby dating the Indian ruin. This, of course, is exactly what he eventually did. In time he moved from Flagstaff to the University of Arizona at Tucson. There he was able to establish the LTRR in 1937, moving his work with tree rings from an avocation to a vocation. Though there are currently a number of centers of tree-ring work in various parts of the world, it seems safe to say that the largest percentage of tree-ring work ever done has been done at one institution—the LTRR at the University of Arizona.

TYPES OF CHRONOLOGIES

A series of tree-ring data to which dates have been assigned is called a time chronology of tree rings—or simply, a chronology. Even though Douglass and associates noted similarities in patterns among trees of a given species at a given site, they realized that there was nevertheless considerable variation from tree to tree. Thus, it made sense to merge data (by averaging) from numerous trees to represent a given species at a given site. This was done after first converting the data to indices rather than merging the raw ring widths themselves. Indices were initially calculated by dividing the ring width of each year by the value for the same year from a curve that was hand fitted to the ring-width series. Merging tends to accentuate those parts of the record that are common among trees and to minimize parts that are not in agreement. The merged data comprise a chronology of tree rings from a set (or collection) of trees, again, of a given species at a given site. The tree-ring series of a single sample radius, such as from an increment core or a transverse section of a tree stem (trunk) may be referred to as a core chronology or a radius chronology. The mean of all radius chronologies (two or more) from a single tree is a mean tree chronology, while the mean of all data from a collection of trees is a mean collection chronology. It has become such a standard procedure to use only mean collections of tree-ring data that the term chronology is used without modifiers to imply mean collection chronology.

The study of tree rings is often referred to as dendrochronology. As dendrochronology has become more popular, the *dendro* prefix (of or pertaining to a tree) is sometimes combined with the field to which tree-ring studies are being applied. Thus, as two examples, dendroclimatology refers to the use of tree rings in climatological problems, and dendroecology refers to the use of tree rings as a tool in ecological studies.

H. C. FRITTS AND DENDROCLIMATOLOGY

In the early years, much of the effort of the LTRR was directed toward chronology building. This meant finding stands of old living trees from which collections could be made and chronologies could be built. Thus, there was interest in obtaining collections from many locations and in extending the length of tree-ring records back in time. In addition, from the very beginning, samples were taken from Indian artifacts and structures and attempts were made to match, or crossdate, the tree-ring patterns in these artifacts with patterns in dated chronologies from living trees. From time to time, Douglass toyed with his old fascination of relating tree-ring data and sunspot cycles. The general consensus became that tree-ring records contained no real cycles, and probably no real connection to sunspot patterns. Later, however, Val LaMarche and Hal Fritts, using more sophisticated statistical procedures such as power spectrum and cross spectral analyses, proposed relationships between tree rings and sunspot cycles, though no direct link was established.

Because of the emphasis on time sequences and dating, it is not surprising that in addition to Douglass, who was an astronomer, members of the LTRR were archaeologists and geo-chronologists. On the other hand, because tree rings are biological, it is surprising or at least interesting that in the early years the LTRR did not include any biologists. After the LTRR had existed for nearly 50 years, the head of the lab, Brian Bannister, decided that perhaps LTRR should add someone to the staff who had a background in tree growth physiology. Thus, in 1959, the LTRR hired Harold (Hal) Fritts from Eastern Illinois University, who was trained in plant ecology under John Wolf in the Department of Botany and Plant Pathology at The Ohio State University.

Upon joining the LTRR, Fritts first set about determining if there was, in fact, a sound biological basis for tree-ring crossdating. By that time, although tree rings had been used to establish many historical dates from Indian artifacts in the American Southwest, there were still those who doubted the authenticity of dates established by tree rings. It is also interesting to note that some of the early authentication of tree-ring dates was done with carbon-14 (C14) dating. More recently, now that tree-ring dating has been accepted as quite accurate, tree rings have been used to authenticate the accuracy of C14 dating. Early work with C14 dating was based on a constant ratio of C14 to C12, though it was known that the ratio has varied with time. Tree rings have been used to fine-tune the variation of that ratio with time.

Fritts found little difficulty in establishing a sound biological basis for cross-dating, but was not satisfied with the paucity in funding for tree-ring research. He eventually found a potential funding interest for estimating, or reconstructing, data of past climatic conditions. In other branches of science, work was under way to estimate past climatic conditions from clay varves (sediment layers) in ancient lake beds and from ice layers in glaciers—both had been determined to be quasi-annual at best. From a number of studies relating the effects of climatic conditions on radial growth, several of which were Fritts' own, it became clear that tree rings contained substantial useful climatic information, even though at the time there had never been a concerted effort aimed solely to reconstruct climatic data from tree rings. It was as a direct result of Fritts' work that significant funding was directed to the LTRR to develop a branch of tree-ring research that quickly became known as dendroclimatology.

Reconstructing climate from tree rings is a major part of dendrochronology, but this text will only brush it lightly since we instead emphasize ecological applications of tree rings. Because many aspects of dendrochronology use methods based on work started at the Arizona LTRR, we will make many references to Arizona tree-ring work. The basic primer on dendroclimatology still remains Hal Fritts' book, *Tree Rings and Climate*. Understandably, advances have been made in the decades since Fritts' book was published in 1976, but it remains the premier reference for those who are serious about reconstructing climate from tree rings. More recently a book by James Speer, *Fundamentals of Tree-Ring Research*, presents the basics of dendrochronology in a very readable format.

The term *dendrochronology* was originally intended to refer to studies involving dating of and with tree rings. In a broader sense, it now often refers to tree-ring studies in general. While the LTRR at Arizona is well recognized as the lead institution in the development of dendrochronology as a science, many other groups have made contributions.

TREE RINGS

Identification of Tree Rings

Botanically speaking, trees are complete plants. Complete plants are composed of six parts: leaves, stems, and roots, along with flowers, fruits, and seeds. Trees become larger by adding new tissue under the bark around all stems and roots. New tissues are derived from cells that are cut off from a cell-dividing region known as a meristem. The meristem that surrounds stems and roots (under the bark) is referred to as a lateral meristem and gives rise to cells that result in growth in diameter. The meristem at a stem and root tip is referred to as the terminal meristem and gives rise to cells that result in growth in length.

The lateral meristem is known also as the vascular cambium or as the cambial layer. Cells that arise from (are cut off from) the inner portion of the cambial layer enlarge and differentiate into xylem, the tissue that we simply refer to as wood (Figure 1.1). Cells that arise from the outer portion of the cambial layer enlarge and differentiate into phloem tissue that we can refer to as inner bark. As the stem grows, previously formed bark can no longer reach around an increasing stem diameter. Additional cell divisions in the bark allow the outer bark to keep pace with the inner bark and cambium.

Examination of a transverse surface, such as the end of a log or the top of a stump or post, reveals what superficially appear as circular lines (Figure 1.2). The outer side of a circular line is the boundary between two annual rings. What appears as a boundary line is brought about by the contrast between small, thick-walled cells at the end of a season and large, thin-walled cells at the beginning of the next season. An annual ring, or tree ring, is not the line itself, but is the area between consecutive boundaries.

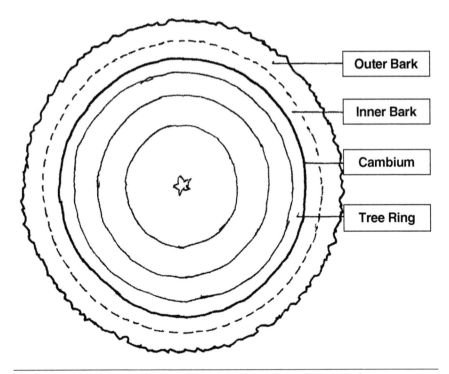

Outer Bark

Inner Bark

Cambium

Tree Ring

Figure 1.1 Sketch showing position of cambium (lateral meristem).

Figure 1.2 Transverse section of pitch pine (*Pinus rigida* Mill.) stem from Cedarville State Forest in southern Maryland. Note the light and dark color contrast that results in the appearance of rings. An individual tree ring extends from the outside of one dark line to the outside of the next dark line. Outside date of this particular sample is 1965.

Microscopic examination of a transverse section allows distinction of a number of cell types. The wood (xylem) of most temperate tree species is composed of tracheary elements, wood fibers, and vascular parenchyma. Tracheary elements are composed of vessel elements and tracheids. Three wood types may be distinguished by their tracheary element types: non-porous woods contain tracheids and no vessel elements (pores); diffuse-porous woods contain small vessel elements throughout the ring; and ring-porous woods contain large vessel elements (pores or water tubes) and small vessel elements. In ring-porous woods the large vessel elements are present only in the early (inner) part of the ring and the small vessel elements are present only in the later (outer) part of the ring.

Non-Porous Wood

Generally, in non-porous wood such as pine (*Pinus* spp.) (Figure 1.3), the circular line distinguishing the ring is the result of tissue that appears darker in color. Color is usually a function of the proportion of a unit area of a transverse surface that is occupied by the secondary walls. Cells formed later in the growth season are smaller and have thick walls (hence, appearing darker) and constitute tissue often referred to as latewood (cf. summerwood). Cells formed early in the growth season are usually large, thin-walled cells (hence, appearing lighter) that constitute what is often referred to as earlywood (cf. springwood). Growth studies have distinguished earlywood and latewood within a given ring on the basis of color, secondary wall thickness, the ratio of radial to tangential dimensions of the cells, or to a combination of these characteristics. The darker color of the cells at the end of one season in contrast with the lighter cells at the beginning of the next season produces the circular line.

Figure 1.3   Enlargement of non-porous pitch pine shown in Figure 1.2. The photograph is a dissecting scope view at about 20x. The outside of the tree ring is to the right.

Ring-Porous Wood

Ring-porous wood is distinguished by a band of very large vessel elements (referred to as water tubes or pores) at the inner portion of the ring. Evidence to suggest that large vessel elements enlarge and differentiate in the early spring from cambial initials that were cut off at the end of the previous growth season will be presented later in this chapter. There is no counterpart in diffuse-porous and non-porous woods for the early portion of the ring-porous wood containing the large vessel elements. Because of this, it might be well to distinguish the early portion of ring-porous rings by referring to it as the pore zone. The portion of the ring-porous ring that is formed immediately after the pore zone is more analogous to the earlywood of diffuse- and non-porous woods than is the pore zone.

The rings of ring-porous woods, such as those of the oak species (*Quercus* spp.) (Figure 1.4), may appear bounded by very thin, light lines rather than wider, darker lines as in many non-porous species. Cells formed at the end of the season (outer part of the ring) may not appear appreciably darker than the main

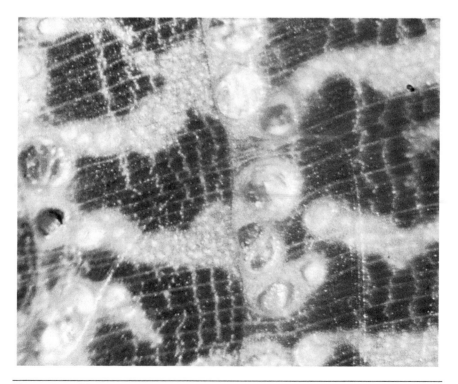

Figure 1.4   Ring-porous woods: dissecting scope view of white oak (*Quercus alba* L.) at about 30×. The outside of the tree ring is to the right.

part of the ring. On the other hand, cells formed at the beginning of the growth season are typically large and thin-walled and hence, appear lighter in color.

Diffuse-Porous Wood

Diffuse-porous woods, such as maple (*Acer* spp.) (Figure 1.5), contain small vessel elements more or less throughout the ring. Because the cell walls formed in the late season by some species are nearly the same thickness as those formed earlier in the season, there may be very little color change across the rings. Therefore, it may be very difficult, or virtually impossible, to distinguish ring boundaries in these woods on the basis of color. The wood of the flowering dogwood tree (*Cornus florida* L.), for example, is quite white with only very subtle anatomical variation noticeable at typical dissecting scope power (10–20×). Thus, some diffuse-porous species, such as flowering dogwood from some habitats, may not be dateable by conventional means. In these cases, alternative techniques such as x-ray densitometry or image analysis might be worth pursuing.

Cambial Activity and Ring Width

We can describe cambial activity as the amount of annual growth relative to the size of the cambium that produced it. Considering a tree trunk in three

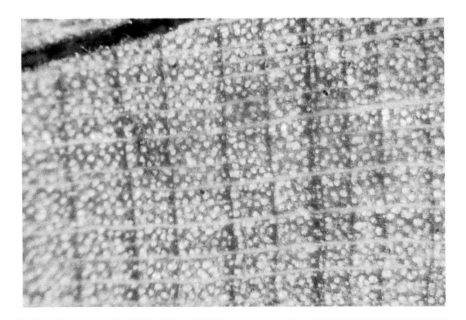

Figure 1.5 Diffuse-porous wood: dissecting scope view of transverse section of the red maple (*Acer rubrum* L.).

dimensions, cambial activity would be the volume of new tissue per unit area of cambium. Cambial activity at a given transverse section could be described as the area of an annual ring relative to the mean circumferential length of the cambium that produced the growth. For purposes of illustration, we can let ring boundaries on a transverse section be represented by perfect circles. This greatly simplifies the geometry. Ring width of the nth ring, w_n, can be described in terms of radius length as:

$$w_n = r_n - r_{n-1} \qquad \text{(Eq. 1.1)}$$

Where r_n is radius length at the outside of the nth ring and r_{n-1} is radius length at the outside of the $n-1$ ring. Thus, circumference would range from:

$$C_{n-1} = \pi 2 r_{n-1}$$

at the inside of the nth ring to:

$$C_n = \pi 2 r_n \qquad \text{(Eq. 1.2)}$$

at the outside of the nth ring. The mean circumference length, μC, which produced the nth annual ring would be:

$$C_n = \pi \left(r_n + r_{n-1} \right).$$

Using the same notations, transverse area of the nth ring, A_n, can be described as:

$$A_n = \pi r_n^2 - \pi r_{n-1}^2$$
$$= \pi \left(r_n^2 - r_{n-1}^2 \right)$$
$$= \pi \left(r_n + r_{n-1} \right)\left(r_n - r_{n-1} \right) \qquad \text{(Eq. 1.3)}$$

Cambial activity, CA, could then be calculated as transverse area, A, relative to the mean circumference length, μC, which produced it:

$$CA = \frac{A_n}{C_n}$$
$$= \frac{\pi \left(r_n + r_{n-1} \right)\left(r_n - r_{n-1} \right)}{\pi \left(r_n + r_{n-1} \right)}$$
$$= r_n - r_{n-1}$$
$$= w_n. \qquad \text{(Eq. 1.4)}$$

Thus, ring width at any particular location can be regarded as an expression of cambial activity at that location. From the observation that, at any given height, the ring width usually decreases with an increasing ring number from the tree center, it follows that the activity of the cambium also decreases. However, as will be discussed in more detail later, though the ring width decreases, the ring area typically increases slightly as photosynthetic capacity of the crown increases. Further, the decreasing trend in cambial activity may at times be reversible, such as during crown or root release.

GEOMETRY OF THREE-DIMENSIONAL TREE RINGS

Work with tree rings deals most commonly with data of one-dimensional (ring width) or two-dimensional (transverse area) components of tree rings. But, of course, tree rings are annual growth components of three-dimensional plants, and are themselves three-dimensional. The intent of the following presentation is to delve a bit into a description of three-dimensional tree rings, thereby placing one- and two-dimensional tree-ring work into perspective.

Each year cambial activity results in a layer of new tissue that surrounds the stems and roots of a tree. The annual layer of new growth is collectively referred to an annual growth layer, or tree ring. The annual growth layer of a tree trunk may be thought of as resembling a sleeve. When viewed on a transverse section, the growth layer appears as a ring. The entire annual growth layer extends out to each stem tip and each root tip. As such, the outer surface of a complete annual growth layer would be the size and shape of the entire tree if the bark were stripped off. Thus, if we could peel off the growth layers one at a time, we could go backwards in time and see what the tree looked like during any of the more recent years of its existence. Of course, the shape of the total tree fades as we continue to go backward in time because we cannot magically see stems and roots that have long since been shed. A canopy-sized forest tree with a lower trunk free of branches may contain few of the branches it had while it was an understory-sized tree.

If we look at tree rings on many transverse sections, we can see that the inner rings are usually wider than the outer rings. One such transverse section is illustrated in Figure 1.6.

The Die-Away Curve

A trend of decreasing ring width with time can be seen in graphs of raw data from a chestnut oak tree (Figure 1.7) and from a loblolly pine (Figure 1.8). This trend of decreasing ring width with increasing age is apparent in sections from all heights in each of the trees.

It may be noted that though data from width sequences at all heights end on the same year (1963 in the oak and 1981 in the pine), the center (pith) ring is on a different year at each height. The influence of climate may be seen by the fact

Figure 1.6   Transverse section from a white oak (*Quercus alba* L.): note the general progression toward smaller ring widths with increasing age.

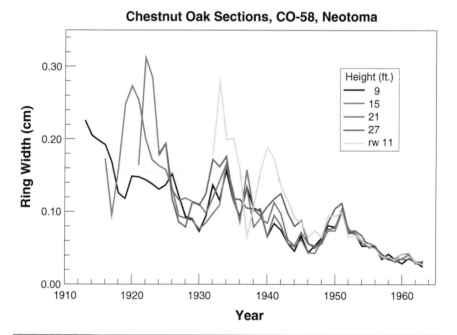

Figure 1.7   Ring width at five different heights in a chestnut oak graphed by year of ring formation. Data graphed in this manner tend to emphasize year-to-year variations. Data are from a sub-canopy tree, CO-58, on the upper northeast-facing slope at Neotoma Valley near Lancaster, Ohio.

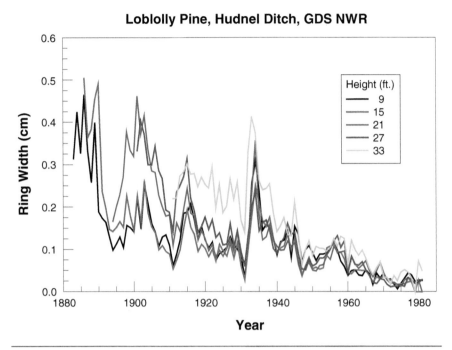

Figure 1.8   Ring width at five different heights in a loblolly pine graphed by year of ring formation. Data are from a Hudnel Ditch site in the Great Dismal Swamp National Wildlife Refuge near Suffolk, Virginia.

that that there is at least a tendency for dips and peaks in the curves to occur on the same years at different heights. For example, the severe drought of the 1930s resulted in poor growth between 1929–1931 in the oak from Neotoma valley in Ohio (Figure 1.7), but was most pronounced in 1931 in the Dismal Swamp pine (Figure 1.8). It may be noted that dips and peaks in the ring widths seem to line up better between heights of the pine (Figure 1.8) than of the oak (Figure 1.7). This may be because the oak was an understory (actually, sub-canopy) tree quite influenced by crowding of surrounding neighbors. The pine, by contrast, was always a part of the overstory.

A better understanding of the shape of the time trend of the ring-width sequence can be achieved by realigning these same ring-width curves by ring number from the pith (Figures 1.9 and 1.10). Because the dips and peaks relating

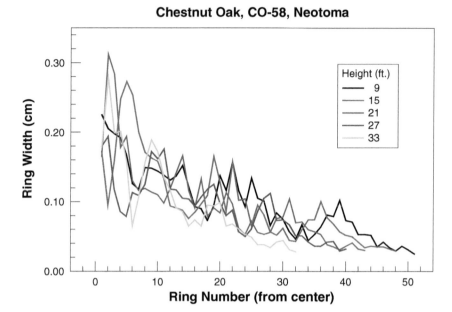

Figure 1.9   Ring width sequences of chestnut oak CO-58 at five different heights by ring number from the pith at each height. Data graphed by ring number tends to emphasize the time trend and to obscure year-to-year variations due to climate. These are the same data graphed in Figure 1.7.

to climate are not matched up by calendar years, the climatic influences are obscured, and the general shape of the time trend becomes more obvious. The time trend illustrated is fairly typical of deciduous forest trees, and is often described as a die-away curve. Many die-away curves appear to approximate the shape of a negative exponential curve. It is not surprising, then, that when mainframe computers came into common usage in the early 1960s, Hal Fritts at the Arizona LTRR wrote a program to mathematically describe the time trend curve of ring-width sequences by fitting either a straight line or a negative exponential curve to the data.

Each height, at any particular ring number in Figures 1.9 or 1.10, is represented by ring growth on a different year. If we examine the sequence from widest to narrowest ring at each ring number, we find that the order varies greatly. Indeed, except for some outer rings in which there is little ring-to-ring variation, the order at any given ring number almost always changes from the previous ring number. This suggests that at any given ring number, the width sequence is controlled more by the climatic effects on growth of each year than

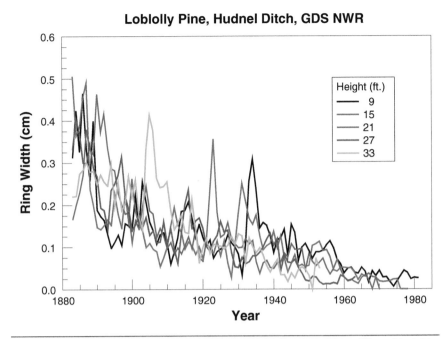

Figure 1.10   Ring-width sequences of Hudnel loblolly pine at five different heights by ring number from the pith. These are the same data graphed in Figure 1.8.

by the height in the tree. This means, for example, that the rings on a 10 cm diameter section taken at the 1 m height of a small tree might look very much the same as a 10 cm diameter section taken from a much greater height in a large tree of the same species.

Profile of the Longitudinal Section

A common model of the way that annual growth layers (tree rings) fit together in a longitudinal section is that of a series of triangles (diagram A, Figure 1.11) in which ring width at the base does not decrease with time (from center outward). If we build into our model the condition that basal ring width decreases with time (diagram B, Figure 1.11), then the width of the center rings decrease with increasing height. Finally, if we also stipulate that the center rings must remain about the same width at all heights, then we arrive at a model in which the profile of the longitudinal section is much more parabolic in appearance (diagram C, Figure 1.11). If the profile of the longitudinal section of each ring is a parabola, then collectively, all the rings in a stem appear as a nested set of parabolas.

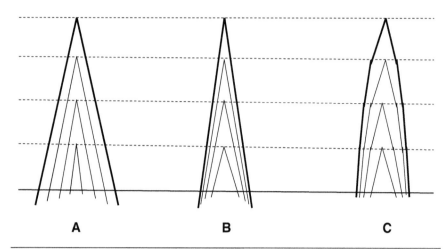

Figure 1.11   Three diagrammatic models of longitudinal sections of a 4-year-old tree. Horizontal lines indicate node heights. (A) This model shows no decrease in ring width with ring number (from center outward). (B) This model shows a decrease in ring width with ring number (time), but shows decreasing width with height for any given ring number. (C) This model holds ring width constant among all node heights at any given ring number and most closely resembles a parabolic shape.

Nested Paraboloids

Rotating a two-dimensional parabola around its longitudinal axis produces a three-dimensional paraboloid. A nested set of paraboloids provides an interesting starting point for a model of how annual growth increments (tree rings) fit together in a tree stem. Figure 1.12 is a diagram of a longitudinal section of a hypothetical stem in which, starting at the center, each succeeding profile (parabola) is simply a longer segment of the same parabola.

By setting annual growth in length to a constant value, some intriguing geometry results. If we let $2r$ = basal width of a parabolic segment, l = segment length, and a = the focal length of the parabola, then the equation:

$$r^2 = 4al \qquad\qquad (\text{Eq. 1.5})$$

describes the relationship between parabola segment length and basal width. If we then designate a constant, s, as the annual increase in segment length, then the relationship for the parabola of the longitudinal section of the outside boundary of the nth year becomes:

$$r_n^2 = 4asn. \qquad\qquad (\text{Eq. 1.6})$$

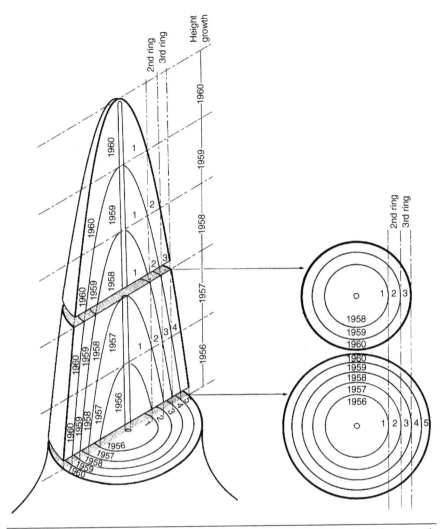

Figure 1.12   Diagram of a longitudinal section of a tree stem in which the annual growth increment profiles are parabolic.

Further, the relationship for the parabola of the inner boundary of the *n*th ring (that is, the outer boundary of the *n*−1 ring) is:

$$r_{n-1}^2 = 4as(n-1).$$

If we assume ring boundaries on the transverse surface describe perfect circles, then the cross-sectional area, *A*, of the *n*th ring is the difference between the area of a circle with radius r_n and the area of a circle with radius r_{n-1}:

$$A_n = \pi r_n^2 - \pi r_{n-1}^2$$

$$= \pi \left(r_n^2 - r_{n-1}^2 \right) \qquad \text{(Eq. 1.7)}$$

Substituting for r^2:

$$A_n = \pi \left[4asn - 4as(n-1) \right]$$

$$= \pi 4as \qquad \text{(Eq. 1.8)}$$

Because π and 4 are constants, and a and s are constants for the hypothetical tree that we are describing, then for that tree, ring area is a constant from ring to ring at any and all heights. Repeating—ring area is a constant from ring to ring at any and all heights. Reflect on that for a bit. It can be thought of as the basic model to describe three-dimensional ring growth.

A long time ago, M. Büsgen and E. Münch in Germany thought that the water conducting portion of a tree trunk was of the same cross-sectional area at all heights in the trunk. They also felt that the overall shape of the trunk was paraboloidal. What they did not do was take it to the final step and describe each tree ring as being of a paraboloid shape and having a constant cross-sectional area at all heights.

In reality we know that the overall shape of a longitudinal section of a tree trunk is not a perfect paraboloid, that ring boundaries on transverse sections are not perfect circles, and that, hence, three-dimensional ring shape is, at best, quasi-paraboloidal. Obviously, because ring shape is the result of growth, and growth is a function of numerous environmental factors that change from season to season and year to year, then ring width, growth in length, and ring shape are by no means constant from year to year; at best they only tend to be. In the real world, perhaps the greatest deviation from the general model of constant ring area is that instead of remaining constant, ring area tends to increase slightly from ring to ring. This will be described and discussed in subsequent chapters.

The Beam of Uniform Resistance to Bending

Let us imagine a beam or pole that is shaped like a tall, slender cone. Let us hold the cone vertically, securing it at the base. If we then apply a horizontal force to the top of the cone until the cone breaks, it will obviously break at the top. The top is the weak point. If we change the shape of the beam to that of a cylinder, and again secure it at the base and apply a horizontal force to the top until it breaks, the break will occur at the secured base. With regard to resistance to bending, the base is the weak point. If the shape of the beam, again secured at the base, is that of a paraboloid, we now have what Büsgen and Münch described as a beam of uniform resistance to bending (Figure 1.13). No one point bends or flexes more readily than any other. There is no weak point. When a horizontal force is applied near the top, there is as much chance that the pole will break at one point as at any other.

In the preceding paragraphs, the shape of a tree trunk was described as being quasi-paraboloidal. Here we hypothesize that a paraboloid is a beam of uniform resistance to bending. Wind may be thought of as a horizontal force. The trunk of a tree offers considerably less resistance to the wind than does the tree crown. If most of the effect of the wind (horizontal force) is on the crown, and the crown is near the top of the paraboloidal trunk, then the effect of wind on the tree acts as a horizontal force applied near the top of the paraboloidal trunk.

In the real world, bending of trees is not quite this simple. The base of a large tree certainly doesn't bend and sway as much as the top. Nevertheless, it

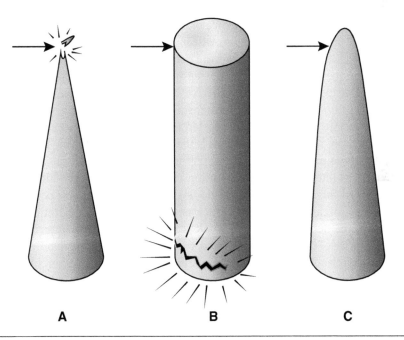

A　　　　　　**B**　　　　　　**C**

Figure 1.13 Vertically held poles attached at the base and to which a horizontal force has been applied near the top. The cone (diagram A) would be expected to break near the point at which the force is applied. The cylinder (diagram B) would break near the attachment point at the base. The paraboloid (diagram C) would have about equal chances of breaking at any point between the point of the force and the point of attachment. If the branch-free portion of a tree trunk is paraboloidal, then the effect of wind on the crown would be analogous to diagram C.

seems more than a coincidence that a tree trunk is quasi-paraboloidal and that a paraboloid is a beam of uniform resistance to bending. If it is more than just coincidence, then there must be some sort of cause-and-effect relationship here. Let's examine this a bit further.

Constriction at the Base of a Discontinuous Ring

It has been stated that a paraboloid is a beam of uniform resistance to bending. It is interesting that trees subjected to the least amount of wind, such as in protected sites and in the middle of dense stands, tend to be tall and have narrow crowns. Trees in exposed positions; that is, in more open grown positions, tend to be shorter and have wide crowns. Trees in the open allocate a smaller proportion of their energy to height growth than do forest interior trees, most likely because more direct solar radiation is available to a greater proportion of the crown of open grown trees.

The width of an individual ring followed from the top of the tree to the ground looks typically like a sequence of rings from the center outward at any given height. In both sequences, growth increases from the top to about midcrown in deciduous trees and from the center outward for several rings, and then both sequences decline with decreasing height and increasing age (ring number). Because growth is not always taking place simultaneously at all heights, and climate does not remain constant, it is not surprising that there are many aberrations to the general, expected trend in ring width of a single ring. One of the more interesting variations is that of an incomplete ring that is present in the upper part of the tree but absent in the lower part.

Figure 1.14 illustrates a vertical section of a series of rings in which the second and third rings from the outside do not form on the right side near ground level. Because of the horizontal to vertical exaggeration of the illustration, a constriction in the outline of the tree trunk is apparent. Thinking in terms of the parabolic beam of resistance to bending, the constriction is a weak point. There would be a tendency for the paraboloid to flex more at the weak point. In the real world, how much of a constriction would be necessary to result in extra flexing at any given point?

Examination of width variation with height of actual individual rings in suppressed red maple and chestnut oak trees revealed occasional rings in which there appeared to be excessive growth in a lower portion of a tree. When tree profile was examined, it was not unusual for excessive growth to occur at the height of an existing constriction. Though the excessive growth might not occur for several years after the appearance of the constriction, the effect of reducing the constriction was apparent. The effect of the excessive growth compensating for a constriction seemed to be more than just a coincidence. In the real world,

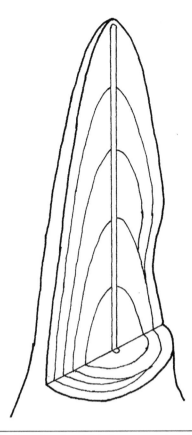

Figure 1.14   Longitudinal section in which the second and third rings from the outside did not reach the bottom of the tree on the right side. The profile of the tree shows a bit of a constriction in the parabolic shape just below the lower termination of the ring. Such constrictions are usually only apparent when the tree profile is graphed using considerable horizontal exaggeration.

we might speculate that sufficient flexibility to result in paraboloid *adjustment* occurs primarily in smaller trunks in which more flexing might be expected. Again, there appears to be ample opportunity for additional research.

SOME INTERESTING VARIATIONS IN GROWTH

Foresters in Austria noted an individual tree growing well above an otherwise uniform plantation of even-aged pines. They found that the tall tree was lodged in the branches of a dead tree that had been left when the plantation was established. The live tree was wedged so tightly in the branches of the dead tree that it could not sway in the wind. When they removed the dead tree, the live tree could no longer support itself and thereby broke off and fell over. Apparently, some of the energy that in other trees was allocated to diameter growth was here allocated to height growth.

In a completely different study, researchers attached a rope and pulley to a large branch that, in effect, removed the weight of the branch from a tree. The rope and pulley were left in place for a number of years. When the rope and pulley were removed, the branch below the point of attachment of the rope wasn't strong enough to support the branch and it fell off. It would appear that the branch was able to move (sway in the wind) only beyond the point of attachment of the rope. Radial growth between the attachment point and the main trunk had slowed significantly, or stopped entirely, resulting in the inability to support the limb without the aid of the rope.

Researchers in Australia guyed several trees so they were prevented from swaying in the wind; that is, the only movement of the trunk was above the attachment point of the guy wires. They also attached guy wires to control trees in which the wires remained loose. When the guy wires were removed several years later, the guyed trees were noticeably taller than nearby similar-aged trees that were not guyed. Unlike the plantation tree that fell after the dead tree that had been propping it up was removed, the guyed trees remained standing after the wires were removed. However, during a substantial wind, the guyed trees broke off below the point of attachment of the wires. No trunks of the control tree that had been wired, but not guyed, broke off. When the researchers sectioned the trees, they found that in the broken trees, no rings had been formed below the point of attachment of the wires during the previous seven years, but there had been excessive radial growth just above the point of wire attachment. Again, when the trunk was prevented from swaying, growth in diameter was reduced in the part of the trunk held rigid, and longitudinal growth was increased in the part above the point of attachment.

Sieve Tube Blockage Hypothesis

Excessive radial growth might be expected just above the height at which a tree is girdled. Could it be possible that the effect of girdling could be produced by sieve tube damage caused by excessive wind inducing extra flexing at the point of greatest trunk constriction? Just a minute disturbance to a sieve plate might be enough to impede the flow of sugars enough to allow a little more growth. If excessive growth did not occur for several years after formation of a constriction, perhaps winds at the time of sieve plate formation were not great enough to result in any damage to plate formation.

The cases in which a branch or trunks were prevented from swaying with the wind could also be explained by the phloem blockage hypothesis. The greatest amount of stem flexing would be expected at or just above the point at which the stem was being held. Growth at or just above that point would tend to be increased and growth below that point would be reduced. An analogy might be the formation of the root collar in free-standing trees.

The Timing of Ring Formation

Tree rings are formed as the result of enlargement and differentiation of new cells produced by the vascular cambium. Growth takes place throughout much of late spring through summer, but the rate of growth on a daily basis varies with time throughout the season, location in the tree, and with both regional climate and local microclimate. An understanding of the timing of growth is important in interpreting environmental information from tree rings. For example, we have been able to use tree rings to date spring and summer floods to within two weeks by identifying flood-induced growth anomalies and then matching their locations in the ring with the estimated time at which that portion of the ring was formed. This is described in more detail in Chapter 5.

Inferences regarding the timing of radial growth at breast height (1.4 m) may be deduced from dendrometer or dendrograph data. Dendrometers measure the change in radius length between the anchor points of brass screws placed in the tree and a measurement point on the outside of the tree by the use of a dial micrometer. Dates of initiation of radial increase for a number of species at the Neotoma Valley research area in southeastern Ohio were determined from weekly readings using Daubenmire-type dendrometers. The dates of the first radial increase in the spring may represent initiation of cambial activity, though radial increase can also result from things other than cambial activity. In other words, it may be necessary to distinguish between radial increase and cambial activity. In addition to spring season cambial activity, two of the most common causes of the first radial increase in spring are (1) rehydration of tissue and (2) changes in temperature. The effects of temperature changes on radial charge are strongest when the tissue temperatures are below freezing. During the growing season when temperatures are above freezing, radial expansions or contractions due to temperature changes are usually slight and therefore difficult to substantiate. Thus, when radial increase in the spring equals the radial maximum of the previous season, rehydration is assumed to be complete, and further radial increase is attributed to cell enlargement.

On a sunny day during the growing season, water can be lost by transpiration faster than it can be taken up by the roots. This results in shrinkage of the outer, younger cells that have not yet fully lignified. Then, during the night, the cells rehydrate (Figure 1.15). If radial growth is occurring, then the radius each morning has increased from the previous maximum.

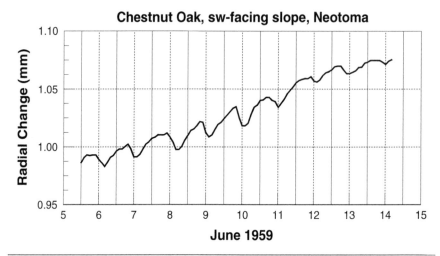

Figure 1.15   Radial change of a chestnut oak on the southwest-facing slope at the Neotoma valley research area, Hocking Co., Ohio. Data were obtained with a Fritts designed dendrograph. Note the diurnal variation composed of radius decreases during the day when transpiration exceeds water uptake, and then rehydration at night. An overcast or rainy day can result in little or no radial decrease. Fritts (1958) defined daily growth from the dendrograph records as an increase from the previous maximum.

Cambial activity starts in the spring, at or near stem tips at the time of initial bud activity (bud swelling and breaking) and then proceeds basipetally. Dendrometer data from Neotoma showed that in large diffuse-porous trees the basipetal progression of initiation of radial increase may take three or more weeks, while initial radial increase occurs more or less simultaneously throughout ring-porous trees.

Bimodal Growth of Ring-Porous Species

Dendrometer and/or dendrograph studies have shown that, in non-porous and diffuse-porous species, growth rate increases rapidly after initiation in early spring, and then slowly decreases after a maximum in late spring or early summer, ending in late summer or early fall (Figure 1.16). Considering that growth initiation may occur several weeks later at the base than at the top of a large tree, it seems reasonable that there should also be a several week lag in growth peak between the apex and base of the tree.

Radial growth rate in ring-porous species is also greatest in the early part of the growth season except that growth appears to be bimodal; that is, there are usually two growth maxima (Figure 1.16). The second of the two maxima

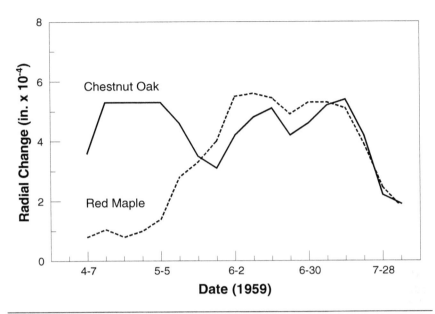

Figure 1.16 Growth rates (as radial change) of a red maple and a chestnut oak on the southwest-facing slope (along with precipitation), from early April 1959, Neotoma Valley research area, Hocking Co., Ohio. The first growth peak of the oak (April and May) is thought to be the result of cell expansion in the pore zone. The second peak in the oak coincides with the single peak in the maple in early to midsummer.

in ring-porous species appears to be time-synchronous with the single peak of diffuse-porous species. The amount of growth indicated by the first peak matches quite nicely with the amount of growth represented by the pore zone. Thus, it is generally accepted that the first peak represents growth related to the formation of the pore zone in ring-porous species.

It is generally agreed that growth in ring-porous species starts simultaneously at all heights in the tree. Recall the contention that the second peak is analogous to the single peak in diffuse-porous species, in that it occurs later in the lower part of the trunk. From this it might be implied that the two peaks may occur almost simultaneously at the top of the tree, but that there may be a several week lag between peaks at the base of a tall tree. Thus, the disagreement among authors concerning the length of time between the two peaks (growth rate maxima) in the ring-porous growth curve may simply be that the trees that were measured were not the same height.

Cambial activity in diffuse-porous species appears to be initiated by auxins and other growth regulators produced at the time of bud break and that emanate from stem tips and then move slowly basipetally. Auxin initiation of cambial activity in ring-porous species is more difficult to accept as the cause of radial increase that occurs simultaneously throughout the trunk.

The Pore Zone Hypothesis

It has been hypothesized that the first peak in the ring-porous growth rate curve does not represent current cambial activity but simply results from the enlargement and differentiation of cells previously cut off from the cambium. That is, the last cells cut off from the cambium at the end of the growth season do not enlarge and differentiate until the following spring.

The width of the pore zone seems to correlate well with conditions at the end of the previous growth season and conditions at the beginning of the present growth season. It could be speculated that conditions at the end of the growth season may control the number of cells in the pore zone, whereas current early spring conditions may control pore zone cell diameters.

The most outstanding feature of the pore zone is the presence of large vessel elements (pores or water tubes). In the main trunk of a large tree, the pore zone of the outer rings is composed almost entirely of water tubes. Near the top of the tree (including the center rings at any height in the tree) the water tubes typically are more scattered, or at least not as densely packed. Could it be that the water tubes are the only cells cut off from the cambium at the end of the previous season? In other words, the tracheids, fibers, and parenchyma cells within the pore zone may result from cambial activity in the spring. If cambial activity in ring-porous species proceeds from the top of the tree, then new cambial initials might be cut off from the cambium near the top of the tree at about the same time as last fall's cambial initials are enlarging into large tracheae. Thus, the fibers end up intermingled with the pores. Farther down in the main trunk, the large pores have already differentiated before current spring cambial activity gets started and the pore zone may be composed only of large vessel elements (pores).

In separately measuring pore zone and latewood in oaks, a pore zone is usually not observed in the center ring. That is, the center ring is composed of pith and latewood, while all subsequent rings are composed of a pore zone and latewood. Because the shoot in which the center ring was formed did not exist at the end of the previous season, there were no cells from which a pore zone could be formed.

Interestingly, the center ring in some wood samples contains something that might be referred to as an attenuated pore zone. The pores in the attenuated pore zone are larger than those typically found in latewood and are smaller than those in the pore zone of subsequent rings. Could it be that these wood samples were taken barely above a node, such that there were a few cells in the unopened bud that were analogous to the cells that normally form a pore zone?

Bob Zahner at Clemson (personal communication) has observed sections of oak trunks cut in winter before bud break or pore zone formation that formed a pore zone soon after being brought into the warmer temperatures of a laboratory. Obviously, the pore zone in these samples could not have resulted from auxins produced at the time of bud break because they were removed before bud break. If the pore zone hypothesis is correct, then the pore zone formed in Zahner's trunk sections were apparently formed from cambial initials present in the sections at the time they were removed from the tree. That is, the cambial initials simply enlarged and differentiated into pore zone vessel elements when the required temperature conditions were met.

Dendrometer studies have also provided information supporting the pore zone hypothesis. Dendrometer screws were installed in a number of trees at Neotoma Valley during the winter dormant period prior to the start of the following growing season of 1955. There appeared to be excessive growth attributed to wounding when the screws were first installed. The excessive growth is apparent in diffuse-porous species of the ring in which the screws were installed. In the ring-porous species, excessive growth occurred in only the latewood of the first year. In the pore zone, on the other hand, excessive growth did not appear in the first year, but rather in the second year (Figure 1.17).

In keeping with the hypothesis regarding pore zone formation, cells that would form the pore zone in 1955 had already been cut off from the cambium at the time the screws were installed. Further, it appears that at the end of the 1955 season when the cambial initials were cut off that would form the 1956 pore zone, enough wound hormones were still present to result in a greater number of cells.

The observations and comments just described suggest that the pore zone in ring-porous species does, indeed, have no counterpart in diffuse-porous species. That is to say, the pore zone in ring-porous species is not analogous to earlywood in diffuse-porous species. For that reason, as pointed out earlier, it would seem preferable to distinguish it from earlywood by referring to it by a different term; that is, the pore zone.

Walnut (*Juglans* spp.) and some hickories (*Carya* spp.) are often referred to as semi-ring-porous. One year, while trying to resolve a problem that I was having with some electric dendrographs, I installed two of the dendrographs on a black walnut (*J. nigra* L.) in my backyard. That allowed me to check on the instruments more frequently than if they were at a more remote location. One dendrograph was placed at about breast height, and the other was several meters higher in the tree. I expected the tree to act as any other ring-porous species; that is, both dendrographs should have shown radial increases at about the same time, and that specific time should have been when pore zone cells were enlarging and differentiating. I further expected both dendrographs to show a second

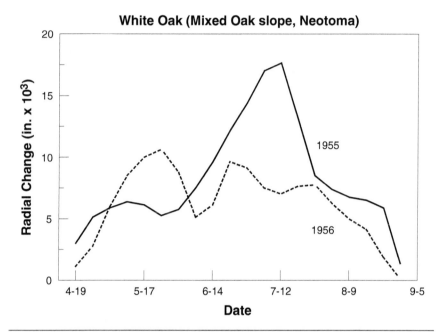

Figure 1.17   Dendrometer data for a white oak on the southwest-facing slope, 1955 and 1956, Neotoma Valley research, Hocking Co., Ohio. Data have been smoothed with a digital filter (0.25 + 0.50 + 0.25). Wound effect from the installation of dendrometer screws before the 1955 growth season appears to have caused excessive growth in 1955 latewood (after about June 1) and 1956 pore zone (before about June 1). The pore zone of 1955 and latewood of 1956 appear more normal. Compare with growth curves from trees in which screws had been installed for four years (Figure 1.16).

peak in radial increase with the upper dendrograph peaking a little earlier than the lower one. To my surprise, the initial radial increase did not occur at the same time at both instrument heights. The upper dendrograph showed radial increase before the lower one, and both instruments described a single-peaked growth curve for the season. In other words, the response that I was seeing was the same as what I would normally see in a diffuse-porous species.

This was one case, in one black walnut tree. Because of the timing of when I installed the dendrographs, did I simply miss the pore zone radial increase? On the other hand, was I simply picking up differences in the effects of wounding as it was associated with the recently installed dendrographs? I don't know. Assuming that the instruments were installed in time to catch the radial increase associated with enlargement of the pore zone, then my results (that must be regarded as tentative, at best) would seem to suggest that formation of the pore zone in at least one semi-ring-porous species is simply part of a single growth peak in radial increase. Clearly, additional work needs to be done before we fully understand pore zone formation.

Initiation of Cambial Activity

Let me dredge up one last story from the dim, distant past. At one time I wished to test my hypothesis that while sieve tubes were being formed, wind-induced flexing of the main stem might cause enough damage to partially block the tubes, thereby having the effect of a partial girdle. My intent was to mechanically flex some small trees at the time of sieve tube enlargement and differentiation.

Trees 1 to 1.5 meters tall from Neotoma research valley were transplanted during the first week of January. We chose a ring-porous species—chestnut oak (*Quercus prinus* L.), a diffuse-porous species—yellow poplar (*Liriodendron tulipifera* L.), and a non-porous species—eastern hemlock (*Tsuga canadensis* (L.) Carr.). We transplanted eight of each species into the plant ecology greenhouse at Ohio State and the same number into a common plot created in the field near the instrument shack at Neotoma Valley research.

Air temperature and solar radiation, among other variables, were measured at both transplant locations. One species (we'll call it species *A*) started leafing out in the greenhouse very soon after being transplanted. A second species (species *B*) leafed out at an intermediate time, well after being transplanted, but well before the same species in the field plot. The third species (species *C*) leafed out in the greenhouse at very nearly the same time as it did in the field plot. Unfortunately, none of the notes or other data of the work were preserved. I could guess which species leafed out at what time, but I do not remember; so, for our purposes here, we will just refer to the species as *A*, *B*, and *C*.

We inferred, with regard to the timing of growth initiation and leafing out, that if any winter conditioning was required by species *A*, it had already been met by the New Year; thus, as soon as it was exposed to the warmer temperatures of the greenhouse, it leafed out. Further, we inferred that growth initiation of species *C* (that leafed out at about the same time in the greenhouse and outside) was controlled by day length; thus, temperature differences between the greenhouse and in the field seemed to have very little effect on timing of growth initiation.

The intermediate species (species *B*) was the challenge. Jerry Koch, who had helped with the whole affair, hypothesized that leafing out may have occurred after some critical amount of heat energy had been received by the trees. Examination of the solar energy data did not reveal an obvious relationship. He then tried summing bi-hourly temperature data above 45°F (~7°C). The sums that he got for the greenhouse trees and the field trees from the time of transplanting to the time of leafing out were quite large, but were remarkably close. This would support a hypothesis that after transplanting, species *B* in the greenhouse and species *B* in the field had received about the same amount of heat energy before leaf bud break.

These results suggest that the timing of growth initiation of some species (such as species *A* and *B*) might vary from year to year depending on temperature conditions. On the other hand, growth initiation of other species (such as species *C*), being more dependent on some aspect of day length, might be expected to be more consistent from year to year. Interestingly, these results could suggest that for the foreseeable future, global warming is likely to affect the timing of growth initiation of some species (such as *A* and *B*), but will have little effect on others (such as *C*). Again, this simple, preliminary study is just the beginning. Additional work is needed before we can accept or reject any conclusions that were suggested by the preliminary study.

SELECTED REFERENCES

Büsgen, M. and E. Münch. (1929). *The Structure and Life of Forest Trees* (English translation). John Wiley & Sons, New York, NY.

Fritts, H. C. (1976). *Tree Rings and Climate.* Academic Press, New York, NY. p. 567.

————. (2001). *Tree Rings and Climate.* Blackburn Press, Caldwell, NJ. p. 567.

LaMarche, V. C., Jr. and H. C. Fritts. (1972). "Tree rings and sunspot numbers." *Tree-Ring Bulletin* 32: 19–33.

Little, Elbert L., Jr. (1979). "Checklist of United States Trees." *Agricultural Handbook No. 541.* U.S. Forest Service, Washington, D.C.

Phipps, R. L. (1967). "Annual growth of suppressed chestnut oak and red maple, a basis for hydrologic inference." U.S. Geological Survey Professional Paper 485-C. U.S. Government Printing Office, Washington, D.C.

Speer, J. H. (2010). *Fundamentals of Tree-Ring Research.* The University of Arizona Press, Tucson, AZ. p. 333.

2

COLLECTING TREE-RING MATERIAL

This chapter will first examine the different types of material used in tree-ring studies. We will then describe the increment borer used to collect tree-ring material, how to collect the material, and finally, a description of how to collect scar samples, root samples, and archaeological samples.

TYPES OF MATERIAL

Any material containing growth layers of trees, and from which a transverse section can be obtained, can be used for tree-ring study. Tree-ring study may, or may not, involve measuring the rings. Trees in temperate parts of the world generally form rings produced during an annual growth period followed by an annual period of growth dormancy. In some parts of the world, trees may or may not produce rings (i.e., distinguishable growth layers) on an annual basis.

For example, the growth layers of a sapote (*Capparis angulata*) from the Sechura Desert in northwest Peru (4° S. Lat.) apparently never completely stop growing during the year. It appears that these trees will periodically put on spurts of rapid growth, probably in response to rain events. The alternation of spurts and non-spurts appear as rings, even though periods of slow growth may last from one to several years in length. Thus, the sapote trees contain rings, but they are by no means annual.

At the edge of the Sechura Desert, in the foothills to the Andes, we sampled other trees, most notably the hualtaco (*Loxopterigium huasango*) that appeared to have growth rings more suggestive of temperate trees. Examination of local climate records indicated that there was one, and only one, rainy season each year. Apparently, the growth surges were in response to water availability brought on by the rainy season. At this point, it is not known whether hualtaco growth actually stops each year. Dendrometer studies conducted on both species by Rodolfo Rodríguez at the University of Piura could shed light on the subject.

Wood samples from nonliving material are good sources of tree-ring material, particularly beams from old buildings. These samples are usually obtained with the express purpose of determining the construction dates of the buildings. For example, much of the early tree-ring work done in North America was done to ascertain construction dates of Southwestern Indian pueblos and cliff dwellings. In the East, tree rings have been used, for example, to do such things as confirm the construction date of a cabin originally built for Gen. Grant during the Civil War. Dating of the cabin is described in some detail in Chapter 3.

The location of wood material is sometimes critical in determining whether the material will be preserved under natural conditions. Charcoal is a form of wood that may remain long after uncharred wood has rotted away. Marion Parker, formerly a dendrochronologist with the Geological Survey of Canada, once used tree rings to help solve a murder case in the Northwest Territory of Canada. Parker's tree-ring dating of some charred wood from a campfire site invalidated the alibi used by the dead man's partner.

Grab samples of buried wood in the Great Dismal Swamp on the Virginia/North Carolina border near Suffolk, Virginia, were carbon-14 dated at around 10K years before the present (BP). It seems reasonable that there may be a continuum of ages of the wood buried in the swamp muck (peat) from 10K years ago to wood buried within, perhaps, the last century. The wood seems composed primarily of Atlantic white-cedar (*Chamecyparis thoides*) and bald-cypress (*Taxodium distichum*). The Great Dismal Swamp could very likely hold a potential for a continuous eastern U.S. tree-ring chronology of 10K years, thereby providing a continuous tree-ring record for the Quaternary.

A little more down-to-earth example is that of George Washington's home at Mt. Vernon in Virginia. Historians have detailed records after George Washington took possession of Mt. Vernon, but they don't know the construction date of the original part of the mansion. An examination of the structural beams beneath the mansion indicated that they are composed of oak (*Quercus* spp.) and American chestnut (*Castanea dentate* (Marsh.) Borkh.). When the beams were examined prior to the 1976 bicentennial, our lab, the United States Geological Survey Laboratory of Tree-Ring Research (USGS LTRR), did not have any local chronologies old enough to crossdate with the beams.

SAMPLING EQUIPMENT

Almost any type of woodworking or woodcutting tool can be used to obtain some form or type of wood sample. A chainsaw or hand bowsaw can be used to remove a small cross section, or transverse section, for tree-ring analysis. A fine coping saw might be used to remove a sample from a wooden picture frame. A wood chisel might be used to carefully remove a wood sample from a living tree that would include the cambial layer. A pruning saw is useful for sampling root material.

Increment Borers

Perhaps the most common form of material used in tree-ring studies is that of increment cores. Increment cores are usually taken from living trees with a hand-operated hollow bit called an increment borer. Powered increment borers, particularly for larger samples, have been built, but are not common. Cores from the most commonly used borers may be obtained in sizes ranging from 4.3 to 5.15 mm diameter and in lengths from 4 to 32 inches (10 to 80 cm).

If a nail is driven into a tree, the tissue once occupied by the nail is not removed but rather compressed into the surrounding tissue. The larger the diameter of the nail, the greater is the amount of compressed tissue and, within limits, the greater is the pressure of the tissue on the nail. If a hollow nail is driven into a tree, the volume of the tissue compressed is essentially equal to the volume of the nail minus the volume of the tissue that ends up inside the nail. An increment borer is very much like a hollow nail with the additional feature of threads that allow the bit to be screwed into the tree. The tissue that ends up inside the bit constitutes the core that we wish to collect.

An increment borer is composed of three parts: handle, bit, and extractor (Figure 2.1). The bit and extractor are stored in the handle for transport. The bit has a collar at the back of the threads. The collar compresses the tissue surrounding the bit to a diameter roughly equal to that of the outside of the bit threads (Figure 2.2). If, after the collar passed, the tissue immediately sprang back toward the bit shank, there would be no advantage to the collar. But, because the tissue tends to spring back slowly, it is possible to twist the bit in as far as desired, and to remove the core and the bit before the tissue applies full pressure to the bit shank.

If the tissue between the diameter of the core and the outside diameter of the bit was chipped away and removed as the bit was inserted, then there would be little or no tissue compression. Chip-removing power increment borers have been custom built for removing large diameter samples from large trees such as sequoias or giant redwoods. Because there would be only minimal compression, chip removers would be relatively easy to insert into a tree. It should be possible to build a conventional-diameter, hand-operated, chip-removing increment borer. Curiously, except for an experimental model that we built at the USGS LTRR, no one seems to have done so.

Forestry supply houses, such as Forestry Suppliers, Inc. (including the former Ben Meadows Company), normally stock two or more brands of increment

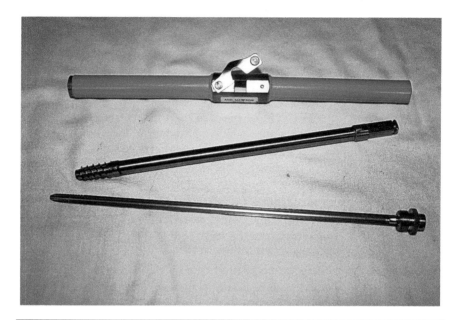

Figure 2.1   Increment borer composed of (from top): handle, hollow bit, and extractor spoon. The bit and extractor are stored in the handle for transport.

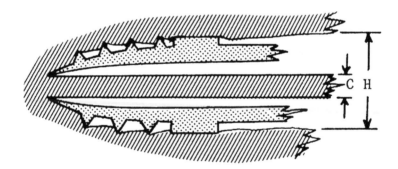

Figure 2.2   Cutting and compression of tissue with increment borer. Compression of wood tissue by borer is indicated by the difference between the size of the hole (H) and the diameter of the core (C).

borers in a variety of lengths and diameters. Generally, the greater the borer diameter and the shorter the length of the handle, the more strength is required to operate the borer. Notwithstanding that longer handles are easier to turn than shorter handles, there is a tendency for the first-time user to purchase a borer that is longer than necessary. The 40 cm (16-inch) borer is the borer of choice by southwestern dendrochronologists at the Arizona LTRR. We at the USGS LTRR found that the 25 or 30 cm (10- or 12-inch) borers work very well for most trees of secondary or tertiary forests in the Eastern deciduous forest.

Most of the older increment borers were built with double threads; that is, a 2-thread bit. More recently, it has been common to build 3-thread models. The pitch of the threads varies among brands and models, but generally, one rotation of a 3-thread model will bore farther into a tree, but will require more effort than one rotation of a 2-thread model. It would seem that the 2-thread bit might be preferable for use in hardwoods. However, some people find that it is harder to start an increment borer than it is to turn it after it has been started, and that the 3-thread borer is easier to start than the 2-thread bit. For this reason, the 3-thread bit may be preferred, even when coring hardwoods.

Archaeological Samplers

The common increment borer sold through forestry supply houses was developed in Scandinavia for use on conifers. The borers available in this country today are imported from Sweden or Finland. They work great on conifers. They will also work on hardwoods, so long as they are kept clean, sharp, and well lubricated with beeswax. They do not work well on a lot of archaeological material such as the very dry wood of structural beams in old buildings. For this reason, various attempts have been made to make borers more suitable to archaeological sampling.

The Arizona LTRR built a sampler from steel tubing, such as the tubing that might be used for automotive brake lines. Teeth were notched in one end, and then the teeth were set. The other end of the tube was notched to fit a hub that could be held by the chuck of a cordless hand drill (Figure 2.3). The set on the saw teeth determined the width of the circular groove cut into the wood. A blow tube was used to remove the sawdust, and a wire with a sharpened hook on the end was used as a parting tool to cut the core loose at the base. Sample tubes were made in a variety of lengths to match the kind of material being sampled.

B. Becker in Germany made an archaeological sampler specifically to remove samples from logs buried in the bottom of the Rhine River. Presumably, these were made to obtain samples from wet wood. The sampler looks like a German plug cutter commonly used in cabinetry except it is larger (Figure 2.4). The sampler cuts a 3 cm diameter hole, leaving a 2 cm diameter core. The sampler is made to take samples up to about 12 cm in length.

The USGS LTRR used the Becker sampler to remove wood cores from dry structural timbers. It worked well, but required quite a bit of WD-40®, an aerosol lubricant, to prevent the sampler from overheating. For a few of the structural timbers, a sample was desired that was somewhat longer than 12 cm. To do this, a sample about 12 cm long was taken; then, inserting the sampler into the

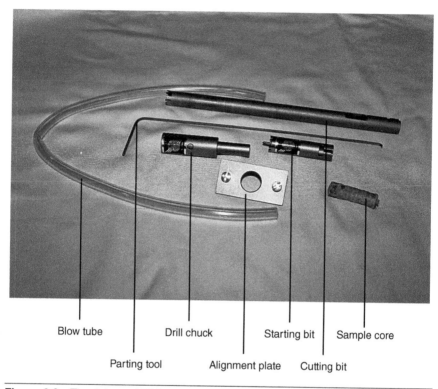

Blow tube   Drill chuck   Starting bit   Sample core

    Parting tool   Alignment plate   Cutting bit

Figure 2.3   The Arizona LTRR archaeological sampler built by Henson.

Figure 2.4   The German archaeological sampler built by B. Becker.

same hole, the shank of the sampler permitted an additional 5 cm coring before the chuck of the power drill came in contact with the beam.

INCREMENT BORER BIT PREPARATION

As with any woodcutting tool, it is important that the increment borer bit is sharp and clean. Perhaps the most common consequences of using bits that are not sharp and clean are twisted and broken cores and cores with surfaces that have not been smoothly cut. Although many authors emphasize the importance of sharpening and cleaning, we have generally been able to maintain sharp, clean increment borer bits without sharpening and with very little cleaning. As discussed in the following section, this is accomplished by using enough lubricant (such as WD-40) to prevent debris buildup inside the bit and by not allowing condensation to form inside the bit as it cools.

Cleaning

Rust results in a roughened surface on which debris can accumulate. Thus, a bit that has accumulated debris almost certainly has some rust and corrosion, often impairing core removal from the bit. Almost as certainly, the cutting edge will have been damaged by corrosion. If damage is not advanced enough to have seriously dulled the cutting edge, the bit can be cleaned and put back into use.

Borer bits can be cleaned with a small bit of toilet tissue that is soaked in WD-40, wrapped around the end of an extractor, and then inserted into the bit. Rubbing the inside of the bit with the tissue cleans and rustproofs the inside of the bit. The same tissue can be used to wipe down and rustproof the outside of the bit including the threads.

Metal Fatigue

Considerable friction develops between the tree tissues and the increment borer bit, particularly in the thread section. No matter what measures are used to reduce friction (discussed later on in the section about preparing the borer), it cannot be totally avoided. Friction causes heat that eventually takes its toll in metal fatigue. Commonly, the first visible signs of fatigue are small, vertical cracks. There may be three or four or more cracks, each extending as much as several millimeters from the cutting edge. Eventually, as cracks lengthen during bit use, a chunk of the bit between adjacent cracks will break out, thus terminating the usefulness of the bit. When a chunk breaks out it may cause a domino effect resulting in the whole end of the bit crumbling away during one twist of the borer handle. No one likes to leave chunks of metal in a tree. Therefore, cease using a bit when cracks are first noted.

Bit Breakage

Because there is always a chance that a bit may break, it is advisable to carry an extra bit when using an increment borer. No matter how carefully increment borers are used, bits do occasionally break. If the bit is relatively new and has been properly used, breakage may be due to a fault in its manufacture. A common fault is that the hole in the bit is not exactly in the center of the bit throughout its length; that is, the walls of the bit are not of uniform thickness. A break resulting from this fault will usually occur about equidistant from each end of the bit, and the hole at the break will be obviously off center.

If a bit is properly cared for to avoid rust, corrosion, and excess heat buildup, it will likely succumb to metal fatigue before the cutting edge becomes very dull. Breakdown by metal fatigue of a properly cared for bit may not occur until a hundred or more cores have been taken.

COLLECTING THE SAMPLE

This section will summarize the steps of collecting an increment core with an increment borer. It is assumed that when you are ready to use an increment core, you have already gone through the process of selecting a collection site, selecting a species to sample, and have selected the first tree to sample. Coring involves preparing the bit, starting the bit, twisting it into the tree, removing the core, and then removing the bit. In addition to describing these steps, we will also cover some special topics, such as removing a stuck bit from a tree, and a trick for hitting the center of the tree.

Preparing the Borer

A problem associated with tree coring is heat buildup caused by friction between the bit and the tree. Because heat buildup induces metal fatigue, it is advisable to reduce friction by the use of a lubricant such as beeswax. Rub the threaded portion of the bit against a block of beeswax to apply the wax to the threads along with some of the adjoining shank of the bit. It is not necessary to apply wax to the entire portion of the bit that is inserted into the tree. As already mentioned, much of the friction is expected to occur in the area of the threads and collar.

The best time to apply wax is when the bit is first taken out of the tree, while the bit is still warm. However, before starting the bit in the first tree, it is important to assure that there is a coat of wax on the bit. It is not necessary that the wax melt into the threads. It is just easier to apply wax to a warm bit than to a cold bit. When coring some very soft non-porous woods, such as Atlantic white cedar (*Chamaecyparis thyoides* (L.) B.S.P.), beeswax may only need to be applied after two or more cores have been taken.

Increment borers are now available with Teflon®-coated bits. The intent is that the Teflon surface will slip through the wood more easily than an uncoated bit, and the coating will reduce the risk of rusting. We have found that for hardwoods, such as oak (*Quercus*), it is still necessary to coat the threads with beeswax, and that the Teflon coating may eventually wear off. Others tell us that the Teflon coating works and holds up quite well when sampling only conifers.

Starting the Bit

It is usually desirable to examine tree rings in transverse sections; that is, to take cores that are parallel to wood rays. This is most likely accomplished by aiming the borer toward the center of the tree, perpendicular to the axis of the trunk. The next step is to simultaneously push the borer into the tree and turn the handle. The most common approach is to hold the borer near the center of the handle, pushing with the palm (Figure 2.5). The other hand may be used to

Figure 2.5 Start the increment borer by a combination of pushing and turning with one hand while steadying the bit with the other hand.

steady the bit to prevent it from wobbling. Wobbling of the bit, before it has been driven far enough to be well anchored in the tree, may cause the core to break. If the break occurs only at the cambium, no harm has been done. However, if the break is just inside the cambium, the outer one or two rings may be very difficult to discern. Wobbling will result in a bent or even a corkscrew-shaped core.

After Starting the Bit

Once the bit has been pushed and twisted enough that the threads are fully anchored into the xylem, the borer will not pull back out of the tree without turning the bit, and there is no more need to continue pushing. The threads will pull the bit into the tree as the borer is turned. At this point, both hands should be used to turn the handle (Figure 2.6).

The borer often squeaks or pops as it is twisted into the tree. These noises result when there is considerable friction between tree tissues and the bit shank. Sometimes no amount of beeswax can completely eliminate the squeaking or popping. Because friction of this magnitude can result in appreciable heat buildup, contributing to metal fatigue, slow the rate of turning the handle so that the popping noise ceases and the squeaking is at least minimized.

Figure 2.6   Pressure need not be applied to push the borer into the wood after the bit threads have engaged the wood.

Continue turning the borer into the tree to obtain the desired core length. Usually, in order to obtain the maximum number of rings, core length should be great enough to at least reach just past the center of the tree. There is no advantage to twisting the bit any farther into the tree than the length of core desired.

Removing the Core

The core is removed from the bit with the extractor, while the bit is still in the tree. The diameter of the bit tube is greater than the core, except near the cutting tip; thus, there is room to insert the extractor spoon. Some people always insert the extractor with the cup of the spoon facing upward so that the spoon slides along the bottom of the tube beneath the core. Others prefer to lightly probe with the extractor to find the side of the tube that is in least contact with the core, and then insert at that position.

It is best to insert the extractor with one steady, continuous motion. If the extractor is pushed only part way in, it may be difficult to restart the extractor and push it in farther. Even if it can be pushed farther, there may be the risk of damaging the core or bending the extractor. As the tip of the extractor reaches the point at which the bit tube narrows, the extractor teeth are forced into the side of the core. This is important. It causes the teeth to grip the core so that it can be withdrawn with the extractor.

The extractor spoon is mounted off center in the knob. Therefore, too much pressure on the knob while inserting the extractor will very likely bend the spoon just below the knob. Essentially all broken extractors that we have seen were broken just below the knob.

After inserting the extractor, break the core loose from the tree before attempting to remove it. This is done by rotating the borer counterclockwise one full revolution while the extractor is still fully inserted. Because the toothed tip of the extractor spoon is tightly wedged between the core and the bit, the core and the extractor will both turn as the borer is turned. If the handle is backed off before the extractor is inserted, the core may not break loose from the tree, and thus will slip through the cutting edge of the bit as the bit is backed off. On the other hand, if the extractor is removed before the handle is backed off, and thus before the core is broken loose, the extractor spoon may slip along the core, failing to remove it, or the core may break somewhere away from the tip, leaving a portion of the core stuck in the bit.

After the handle has been backed off, breaking the core loose from the tree, remove the core from the bit with the extractor. Place a hand beneath the extractor as it is withdrawn to catch any loose pieces that may fall off the extractor spoon. This, of course, is particularly important if the extractor was not inserted beneath the core, but rather to the side or top of the core.

Removing the Borer Bit

After removing the core from the bit, it is very tempting to admire one's handiwork by examining the core while the borer bit is still in the tree. Avoid this temptation, and immediately remove the borer from the tree. As already explained, compressed tissue around the bit slowly springs back to its original shape, increasing pressure on the bit. This is quite simply because the hole cut by the bit is smaller than the diameter of the bit. If too much time elapses while the borer is in the tree, the borer may become stuck. A few minutes may be too long. Removing a stuck bit can be very difficult. Also, as previously mentioned, it is a good idea to apply beeswax to the threaded end of the bit while it is still warm immediately after withdrawing it from the tree.

It appears that corrosion is much more harmful to the cutting edge of the bit than is fresh wood of a living tree. Borer bits and extractors replaced in the handle immediately after the bit has been withdrawn from the tree will almost always rust overnight when sampling in the summer in the humid eastern part of the United States. Moisture that is present inside the bit as a direct consequence of boring is thus trapped inside the handle and cannot quickly evaporate into the air.

Storing the bit in the handle only after it has cooled to ambient temperature helps avoid rust. In winter weather, the bit may not completely cool for 10 to 15 minutes; in summer weather, it may take much longer. At times it may be necessary to store the handle before it has become completely cooled. For example, it may be better to store the bit in the handle than risk bumping the bit into something that could damage the cutting edge. When the bit cannot be allowed to cool completely before storage, a couple of squirts of oil from an aerosol can into the bit tube before placing it in the handle will usually prevent rust formation. Some of our co-workers routinely place a couple of squirts of oil in the tube at the end of a collection session, whether or not the bit has cooled to ambient temperature. If no rust or corrosion is allowed to develop in the bit, it may become stained, but otherwise will remain remarkably clean.

Removing a Stuck Bit

As previously described, failure to remove the borer from the tree immediately after extracting the core from the borer bit commonly causes the borer bit to stick or lodge in the tree. Driving the bit threads past solid wood into rotten tissue or a hollow void also often results in a stuck bit.

Usually, a stuck bit will not have seized in the tree. Indeed, the handle will usually turn more easily than when normally backing the bit out. Whether the bit was left too long unattended or was driven into rotten wood, tissue has sprung back against the borer shaft with enough force that the collar at the back of the threads cannot be easily moved back far enough into the solid wood in order to re-engage the threads.

Often a stuck borer can be restarted by simultaneously pulling and twisting the handle. If the borer is tightly stuck, it may be necessary to perform this operation as a very hard jerk while turning the handle about an eighth of a rotation. Usually, a series of jerk-and-turn operations will back the borer out far enough to re-engage the threads. This procedure is almost always successful in restarting a borer that has encountered a rotten spot. A borer that has become stuck in solid wood because it was left in the tree too long may be more difficult to remove.

Rarely, a borer may become stuck so tightly that no amount of jerking and turning will restart it. In all of our years of working with increment borers, we have only encountered three instances when the borer was stuck too tightly to be removed by the jerk-and-twist procedure. Two of these instances involved leaving the borer in the tree too long while examining particularly interesting cores. Both borers were removed by the use of a small rope. The rope was looped around the borer handle and an adjacent tree, pulled tight, and the two ends tied together. The borer handle was then turned counterclockwise, twisting the rope. Continued twisting tightened the rope enough to pull the borer back enough to allow the threads to engage. It was then possible to remove the rope and back the borer out in the normal fashion. In the third case, the borer became stuck in very soft, rotten wood at the center of the tree. It was not possible to remove the rope until the bit had been completely withdrawn from the tree.

When using the rope method to remove a stuck borer, be sure to stand clear of where the handle and the rope might suddenly fly off. Remember, the only thing holding the handle on the borer bit is a small clip.

Removing a Stuck Core

Occasionally, after removing a borer from the tree, part of a core may be stuck in the bit. If any part of the core is sticking from the end of the bit, do not attempt to push it back into the bit or pull it back into the bit with the extractor. Break it off and then attempt to remove the remaining fragment inside the bit by the use of the extractor.

It is possible to have a very tiny plug or core left in the cutting edge end of the bit that the extractor teeth cannot reach quite far enough to grip. If the plug is not too tightly wedged, it is sometimes possible to remove it by simply starting a new core, and letting the new core push the plug loose. On the other hand, if the plug is tightly wedged it may not be possible to engage the threads to take a new core. In this case, there is little choice but to drive the plug into the bit from the cutting edge end. If an attempt is made to drive the plug out by inserting a wooden dowel into the bit tube, it may only succeed in wedging the plug even tighter. Further, spraying a fluid in the tube may only exacerbate the situation by swelling the core and causing it to become even more tightly wedged. A small, whittled wooden stick seems to be the most practical tool to use to push the plug back into the bit. Use of a whittled stick in the field, as opposed to simply picking up a stick lying nearby, ensures that any sand or grit that could harm the cutting edge has been whittled away. A handy alternative to the whittled stick is a simple golf tee.

Sampling the Center Ring

Some studies require the determination of tree age at sample height. This, of course, necessitates sampling the center ring. It sometimes seems that trees love to play games with the field scientist, particularly if the investigator is intent upon sampling the center ring. Even in symmetrical tree trunks, the botanical center seems to not be at the geometric center. Consequently, one rarely hits the center on the first attempt. Further, the tree must be capable of moving its botanical center between the time of taking the first and second samples. We have no other plausible explanation as to why the center is so often missed on the second attempt. Fortunately, there is a neat trick to greatly improve the chances of hitting the center on the second or third attempt. The trick is simple enough that you may have already discovered it on your own.

When extracting the first core, back the handle off exactly one revolution (360°). This leaves the core oriented in the extractor spoon in the same way it was when attached to the tree. Then, when the core is withdrawn, the arcs of the rings near the center indicate whether the botanical center is to the right or the left of the first core hole along with how far right or left the center is.

Remove the core from the extractor spoon and place it back in the hole from which it came, leaving several centimeters of core sticking out of the tree. For the second attempt, orient the borer parallel to the first core, and the estimated distance to the right or left. If the estimated distance is quite short, it is also a good idea to place the borer a few centimeters up or down from the first hole to make sure that there is good solid wood into which the borer threads can grip. If the tree is not much larger than about 60 cm in diameter, this method will usually yield the center ring on the second or third try.

Plugging the Core Hole

Customarily, the hole that remains in the tree after a core has been removed is not plugged. Although the wound caused by boring can provide an entry point for bacterial and fungal infection, the risk of infection appears to be minimal for most tree species. Occasionally, a landowner will grant permission to core trees on the condition that the core holes be plugged or sealed. A dressing—such as the kind that is commonly used for cut surfaces when branches are pruned—may serve to placate the landowner. I am not convinced that it always does more good than harm. We have noted a number of cases in which plugged holes failed to heal as rapidly as unplugged holes.

HANDLING THE CORE

This section includes field examination, methods of storing cores, the drying of cores, and mounting cores.

Field Examination

Upon removing the core from the tree, it is advisable to perform a cursory field examination. Even casual observation may reveal faults in the core that may render it unusable. For example, inadvertently sampling too close to an adventitious bud or a branch trace of a long-gone branch can yield distorted ring boundaries. It is certainly better to discover the unusable core while in the field than after having returned to the laboratory.

It may be necessary to more closely examine a core in the field. For example, accurate ring counts in the field may have a bearing on further activities of the same field trip. Or, it may be necessary to determine if the core contains a predetermined number of rings.

Field counts of rings made directly on an undisturbed core as it is withdrawn from the tree are rarely reliable. More precise examination in the field can be accomplished by hand-surfacing the core and then examining it under a hand lens. Paring or slicing with a sharp knife can surface the core. It is advisable to hold the core in some sort of clamp during the slicing operation (Figure 2.7). Care should be taken to orient the core in the clamp so that the exposed surface of the core will yield a transverse section of the wood when sliced. This can easily be done by turning the core in the clamp until the grain of the wood (vessels and wood fibers) exposed on the end of the core is parallel to the sides of the clamp.

The exposed surface of the core is then sliced with a sharp knife. I prefer to use a surgical scalpel with a standard #10 disposable blade. For very hard wood, such as some *Quercus* spp., the blade may need to be changed after one or two cores, depending in part on the length of the core being surfaced. A more detailed description of slicing cores with a knife is provided under the heading *Surfacing the Core*.

The surfaced (sliced) core can be examined in the field with a hand lens—a 10× Hastings triplet works well for most purposes. The ring boundaries of some diffuse-porous species may be more easily distinguished after the sliced core

Figure 2.7   Core clamps used to hold cores during surfacing, crossdating, and measurement. Shown are a wooden clamp and an aluminum clamp. An increment core is mounted in the wooden clamp.

has been allowed to air dry for a half hour or more. However, the boundaries of some diffuse-porous species can be extremely difficult, if not impossible, to distinguish under field conditions.

Storing Cores

Many methods of storing and transporting cores under field conditions have been described in the literature. By far the most popular method is to store the cores in soda straws. Paper soda straws are preferable to the more commonly available plastic ones. The paper soda straws are rigid enough to prevent the cores from warping during air drying, yet porous enough to allow the cores to dry without developing profuse crops of mold. Further, when handling small-diameter cores or broken cores that may easily slide out of the straw, the ends of a paper straw may be neatly crimped and closed. Of course, paper straws cannot be used if it is desired to store the core in its original, fresh, wet condition. Cores may be kept fresh in the field by any of a variety of ways including, for example, temporary storage in small ice coolers.

The straw should be labeled with an identifying note or number. The simplest procedure is to use a sequence number that is keyed to field notes containing such information as date, location, site, and tree descriptions. By convention, I number the end of the straw that contains the bark end of the core. Knowing which end of the straw contains the bark end of the core may be helpful when removing the core from the straw.

The paper straws can be field-stored in mailing tubes that have been cut to length. The straws are placed in the tube such that the numbered end of the straw is at the top. These containers have been mailed without damaging the cores.

Much of the coring from secondary or tertiary forests of the eastern decidu-ous forest results in full radius-length cores that easily fit into 25 cm soda straws. Some cores, however, are too long to fit into a single straw. In such cases, *cut* the core and store it in two or more straws. Do not *break* the core. A core, particu-larly of a ring-porous species, will often break along the pore zone. When the core is put back together for ring-width measurement, a certain measurement error is incurred in the one or two rings containing the break. To prevent this, cut the core diagonally across a ring boundary such that a portion of the same boundary remains on each piece. Cut about halfway through the core, parallel to the grain, and then twist the two pieces apart. The first (outer) core piece is placed in the first straw in the usual manner. The second core piece is placed in the second straw with the outer (cut) end at the numbered end of the straw.

Occasionally, if the cores are to be permanently mounted in blocks instead of being stored in straws, it may be better to leave a long core sticking out of a straw in the field than to cut it.

Air Drying Cores

Even if stored in porous paper straws, the drying rate is greatly reduced if the straws are left in the mailing-tube straw containers. When the straws are removed from the tube, the cores often dry enough in a day or two that they can be easily sanded. Additional drying may be desirable if the cores are to be permanently mounted instead of being stored in straws.

Removing Cores from Straws

Cores are easily removed from straws by the use of a small-diameter rod. A 3 mm (⅛-inch) brass-welding rod works quite nicely. To remove the core, the rod is inserted from the unnumbered end of the straw. If the rod is inserted from the numbered end that contains the bark-end of the core, the rod may cause the bark to flake and jam the straw. The rod may be inserted into straws with crimped ends.

Mounting Cores

Most laboratories customarily permanently glue their cores in grooved, wooden mounting blocks. Most of our cores are not mounted. Our cores are removed from the straws and placed in core clamps for surfacing, crossdating, and measurement, and then placed back into the straws for permanent storage. We will mount our cores, however, if (1) the cores are fragile or broken into many pieces, as is common with eastern hemlock (*Tsuga canadensis* (L.) Carr.) or (2) it is expected that the cores will be repeatedly handled. It is always possible to mount a core at a future date, but it is very difficult to remove a core from a mount.

Laboratories that mount their cores usually prefer to air dry the cores before mounting. Fresh cores almost always pull apart in the mounting block while the core and the mounting medium are both drying. The cores must be very dry before mounting in order to avoid the breaks that can make measurement a bit awkward.

There are advantages to mounting only one core per block, rather than the common practice of mounting two to four or more cores side by side on a single block. Some of the subtleties of sanding (surfacing) the cores are more easily accomplished if each block contains only one core. Maneuverability while sanding is further enhanced if the mounting block is narrow enough to include very little block material on either side of the mounted core. A block that is about one centimeter square in end view works very well. One surface of the block can be grooved with a router to accommodate the diameter of the cores being mounted. The groove should be shallow enough to allow about half of the diameter of the core to protrude above the block, thereby allowing a reasonably wide finished (sanded) surface.

It is important to orient the core in the block such that the wood grain (vessels and fibers) will be at right angles to the finished surface. It is thus important that the core is not twisted appreciably. Any glue that is commonly used in cabinetry can be used as a mounting medium. Some means of holding the core in the block should be provided until the glue dries—rubber bands work very nicely.

In working with material in which the ring boundaries are particularly difficult to discern, it may be preferable to hold the cores in clamps rather than in permanent mounts. Sometimes it is found that resurfacing the cores at a slight angle to a more perfect transverse section enhances boundary distinction. Additionally, a clamped core with rings that are particularly difficult to delineate can be turned over, allowing the opposite side to be surfaced.

Twisted Cores

In twisted cores the angle of the wood grain changes along the length of the core. The wood was not twisted while still in the tree, nor does twisting occur as a core dries. Pronounced twisting is caused by the core dragging in a very dirty or rusty borer bit tube, or by a dull, burred, or improperly sharpened cutting edge. However, particularly with some soft conifer woods such as *Tsuga* spp., it is sometimes impossible to avoid minor twisting, even when using clean, sharp borers.

Severe twisting may necessitate some sort of special treatment in order to straighten the core. Simply soaking in plain water may be sufficient. However, it may be necessary to soften the core with live steam. After the core is softened, it may be straightened and held in a core clamp while it dries. Recheck the core from time to time as it dries to ensure that the core has not shrunk enough while drying to loosen itself in the clamp and perhaps partially retwist.

If total twisting is less than about 45°, it may be better to leave the core twisted. In such cases, mount the core so that the approximate midpoint of the core length is oriented correctly and each end is twisted about equal amounts in opposite directions.

Surfacing the Core

Attempts to distinguish ring boundaries on unsurfaced material are generally not reliable. Even in conifers in which most boundaries appear distinct on the unsurfaced core, there is a risk of missing very tiny rings or misidentifying false rings.

Slicing

Slicing or paring of cores with a knife or scalpel is usually not particularly effective unless the material is fresh. Lignified cell walls are not easily cut cleanly once the material has dried. Softening, as with live steam, will generally not yield as clean a cut surface as will fresh material.

The blade used for surfacing must be nearly as sharp as a sliding microtome blade. A good, experienced wood carver may have knives sharp enough to do a respectable job, but this is rare indeed. Some people use craft knives, such as Xacto®, though I prefer surgical scalpels with disposable blades.

Pare off thin layers of wood to produce the desired surface. Cut with a slicing motion. Do not push the blade through the wood. In executing a single slice, the portion of the blade that is in contact with the core should progress from blade heel to toe.

Properly sliced surfaces have an advantage over sanded surfaces in that vessels (angiosperms only) are left open and not filled with sawdust. Ring boundaries of some difficult diffuse-porous material are sometimes more easily discerned on a sliced than on a sanded surface.

The act of slicing cores is almost an art. Do not expect to get high-quality surfaces without considerable practice. Because good sanding techniques are much more easily learned, many people quickly abandon attempts to develop good sliced surfaces.

Sanding

Anyone who has experience in woodworking or furniture refinishing knows that it is relatively easy to smoothly sand the side of a board. But to smoothly sand the end of the board, the *end grain*, is quite a different story. The transverse surface of a wood sample is the *end grain*.

Sanding can be done by hand or by a machine such as a sanding disc on an electric drill. As our sanding techniques evolved at the USGS LTRR, we ended up preferring a sanding disc on a drill press. Use of a drill press leaves both hands free to maneuver the core clamp, and usually provides both higher speeds and a greater range of speeds than an electric hand drill.

The most practical sandpaper grits for use on cores are finer than those normally available in ready-cut sandpaper discs. Therefore, we would buy sheet sandpaper and cut our own discs. We normally stocked grits of 220, 280, 320, 400, 500, and 600. For each core, only two or three grits of sandpaper are typically used—starting with a coarse grit and ending up with 400, 500, or 600 grits.

Each grit should remove the scratches of the previous grit. When first attempting to sand cores, the core should be scanned under a dissecting scope after each step. Scratches that obscure ring and cell boundaries are obvious when viewed at 15 to 30 power of magnification. Sometimes it will be necessary to repeat a previous grit, or to use an intermediate grit.

The objective of sanding is to prepare a surface similar to a cut surface. The purpose of the coarsest sanding is to prepare a flat surface. The purpose of the finer sanding is to reduce and finally remove the burred edges of the cell walls so that they end up being sharply and clearly defined.

As a final step, buff the core with a lamb's wool pad. When viewed under a dissecting scope, the difference between final sanding and buffing is remarkable. Buffing not only does some polishing, but also removes some of the sawdust clogging the vessels. Considerable pressure between core and pad may be applied during buffing. However, if there is concern that the sawdust remaining in the vessels may act as an abrasive during buffing, use the buffing pad to lightly dust the core surface before applying appreciable pressure.

Enhancing Indistinct Ring Boundaries

The literature abounds with methods to enhance identification of indistinct ring boundaries. Many of the notes refer specifically to problems with rings of diffuse-porous species. There is no way that we could have tried every known method, but we have tried many. We have rubbed various types of oils, kerosene, and even perspiration on the surfaced material. We have tried a variety of stains, including some used in techniques for slide preparation. We have tried shoe polish, and we have even tried lightly burning the surface and then resanding.

We have found nothing that is a marked improvement on a properly sanded surface. As already noted, some ring boundaries may be more easily distinguished on a cut surface than on a sanded surface. More often than not, though, we find that we must touch up a cut surface with some *spot* sanding to produce the most *readable* surface.

Even the best surface can be enhanced by lighting and magnification. For the vast majority of wood samples, good quality lighting is important, but almost any type of good quality light will do. For the more rare samples in which lighting is quite critical, there seems to be no best combination of lighting and magnification. Experimentation with the particular specimen at hand seems to be the only answer. Sometimes light at a very low angle will cast shadows in the vessels, emphasizing differences in vessel size. High-intensity light, such as that which can be achieved with fiber-optic lamps, may be helpful, as may various light filters. Sometimes features not found at high magnification will be more obvious at low magnification, and then confirmed at higher magnification. Regardless of these techniques, one of the most important and indispensable tools is a microscope with high resolution capability.

Regardless of preparation techniques, some material, particularly that of some diffuse-porous species, may be difficult or impossible to read. When we first attached a TV camera to our measurement equipment microscope, we attempted to achieve the most realistic picture possible. We then found that by deliberately modifying the color and contrast, we could sometimes more easily distinguish some otherwise obscure anatomical features. The closed-circuit TV thus became another tool for distinguishing ring boundaries.

Techniques for enhancement that work well for one person may not seem the best for someone else. In dealing with particularly difficult material, it is advisable to try a variety of techniques. More than likely, however, you will develop your own favorite techniques.

ADDITIONAL COLLECTING METHODS

The increment borer is the tool of choice for collecting tree-ring material from living stems. On the other hand, increment borers are difficult, or even impossible, to use for some specialized types of collecting. As already mentioned, increment borers are usually not very satisfactory in removing dry wood from structural beams of old buildings.

SAMPLING OLD BUILDINGS

There appears to be a growing interest in dating old buildings in eastern North America. Historical architects have done some remarkable work of dating old buildings using such techniques as dating from paint chips. Usually their dates are collective estimates based on two or more lines of evidence. Often the best that can be achieved is a range of years during which construction most likely occurred. The attraction of tree-ring dating is that, potentially, the onset of construction can be determined to within a year.

The structural beams of colonial buildings were typically made from local trees. A common practice was to cut trees during winter and then use them in construction before the following winter. Thus, we can usually expect that construction was at least started during the year after the date of the last formed ring on wood used for structural beams.

Early structural beams were usually hand hewn to a set rectangular shape. Hewing the logs was quite labor intensive. A practical procedure was to harvest logs so that only a minimum amount of material had to be removed to achieve the rectangular shape. In the process of hewing the log to produce a straight beam, any slight curve in the original log might result in patches being left on the inside of a curve.

The structural timbers are dated by crossdating the ring-width patterns in the beam with an existing, dated pattern. If the sample to be dated was taken from a bark surface, then the date of the last ring formed before the tree was cut can be determined. The usual procedure is to sample from a bark surface toward the center of the beam. There is usually no need to sample any farther than just past the center of the beam.

Samples obtained with the Arizona-type sampler by Henson (Figure 2.3) and the Becker-type sampler (Figure 2.4) produce samples that look like oversized increment borer cores. Perhaps equally common are samples obtained by either sawing the end off of a beam or sawing a narrow V-shaped slice from anywhere along a beam with an edge that contains (or once contained) bark. Often, samples taken from anywhere other than the end of the beam can affect the structural integrity of the beam. These, then, are often taken from beams that have been permanently removed from the structure.

SELECTED REFERENCES

Jozsa, L. (1988). "Increment core sampling techniques for high quality cores." Forintek Canada Corporation, Special Pub. No. SP 30. p. 26.

Phipps, R. L. (1985). "Collecting, preparing, crossdating, and measuring tree increment cores." U.S. Geological Survey Water-Resources Investigations Report (USGS WRIR) 85–4148. http://pubs.usgs.gov/wri/1985/4148/report.pdf.

Speer, J. A. (2010). *Fundamentals of Tree-Ring Research*. University of Arizona Press, Tucson, AZ. p. 333.

Stokes, M. A. and T. L. Smiley. (1968). *An Introduction to Tree-Ring Dating*. University of Chicago Press, Chicago, IL. p. 73.

———. (1996). *An Introduction to Tree-Ring Dating*. University of Arizona Press, Tucson, AZ. p. 73.

3

CROSSDATING, MEASUREMENT, AND STANDARDIZATION

The term dendrochronology can be broken down into three roots: *dendro* relating to trees, *chron* relating to time, and *ology* referring to *the study of*. A simplified definition might be: *dendrochronology* is the science pertaining to dating of and with tree rings. In a more generalized sense, dendrochronology has come to refer to, or be synonymous with, tree-ring studies.

This chapter contains some of the most important basics of dendrochronology. We will be discussing crossdating, tree-ring measurement, and standardization. The crossdating portion of the chapter will discuss some methods of crossdating, and then present an example application. A rather brief description of some of the problems encountered in measuring tree rings will be presented. And, finally, some methods of standardizing tree-ring data will be described.

CROSSDATING

Crossdating is a procedure for assigning absolute dates to tree rings. Other methods of dating, such as carbon-14 (C14) dating, assign a plus-or-minus (±) to the date because they are not exact. For example, a C14 date might appear as 2,200 ± 50 years, meaning that the actual date may not be exactly 2,200 years ago, but is most likely sometime between 2,150 and 2,250 years ago. On the other hand, a correctly crossdated tree-ring series might contain a ring dated as 1637, meaning that the ring was actually formed in the year 1637, not 1636 or 1638. This chapter will explain how and why crossdating works, and will then discuss the methods for doing some actual crossdating.

The Art

A. E. Douglass and his associates at the Laboratory of Tree-Ring Research (LTRR) at the University of Arizona developed and refined crossdating methodologies in the United States. A number of people have been cited in the literature for having *discovered* crossdating, thus laying the foundations for dendrochronology. That notwithstanding, there is little question that the bulk of the credit really should go to Douglass and his associates at the LTRR. When they were finally convinced that crossdating indeed was possible and produced exact dates, they were also convinced that the importance of crossdating should not be taken lightly. A casual or sloppy practitioner could easily assign inaccurate dates, thereby damaging the fledgling new field of dendrochronology. Because of this, those inquiring about how to crossdate were cautioned to fully familiarize themselves with the entire process. The reaction to this was the misconception that Arizona surrounded crossdating with some sort of mystique, and didn't want outsiders to learn how crossdating really worked.

The Methodology

Crossdating is the procedure by which absolute dates are assigned to each tree ring. Basically, crossdating amounts to matching ring-width patterns among wood samples. Using accepted crossdating procedures ensures that anomalies such as missing rings and multiple rings are identified. Correctly crossdated material produces absolute dates for each ring. Simple ring counting does not involve crossdating and, thus, does not necessarily result in absolute dates. Ring counting can provide the same dates as crossdating, but only in situations where the absolute date of at least one ring is known (usually the outside ring), and one and only one ring is present for each year. It is not possible to know how many missing or multiple rings will be found in material of a given species in a given habitat until several collections of material from an area have been studied.

Michael Duever, working in the Okefenokee Swamp in Georgia and later in the Corkscrew Swamp Sanctuary in the Florida Everglades, used to say to never go into a new area with preconceived notions about what will or will not crossdate. Working primarily with baldcypress (*Taxodium distichum* (L.) Rich.), he found some areas in which the trees crossdated easily, some in which the trees crossdated only with great difficulty, and some in which he could not get trees to crossdate at all. When starting to work in a new area, his standard procedure was to cut down some smaller trees. He then worked with transverse sections. He found that examining each ring all the way around the section gave him the best, and quickest, appraisal of discontinuous rings. This, in turn, gave him a pretty good idea of what to expect if he were to use increment cores. In other words, if discontinuous rings were quite common and numerous on a section, then it was a pretty good bet that crossdating increment cores would be challenging at best.

Marking Dates of Rings

Starting with a well-surfaced wood sample, most dendrochronologists find it convenient to mark rings as reference points. It can be exasperating to be 300 rings into measuring a core and then lose your place when the core is accidentally moved or you look up because of a momentary distraction. A common marking system is to use dots: one dot for a decade, two dots for a half century, and three dots for a century. Some people use an ink pen to place dots, while others will make pin pricks in the wood sample (usually an increment core) with a finely pointed dissecting needle. We prefer to use a finely pointed drafting pencil. If dating errors are discovered, it is a simple matter to erase the incorrectly placed dots and replace them in the correct rings. In large rings, it is possible to write a number within the ring boundaries (for example: a 3 could represent 1930). It is also helpful to make some sort of mark or notation indicating the location of missing rings. A common notation is to place a small circle on the ring boundary where the ring is missing (Figure 3.1).

Missing Rings

Cambial activity is initiated in the spring by growth regulators emanating from in or near the bud tips of branches. Because the production of new woods requires energy, growth may be thought of as an expression of food availability. The paraboloidal pole described to represent the main stem of a tree (Figure 1.12) may be thought of as having resulted from an ample food supply every year. The leaves, of course, are the food source for the tree; hence, the food source may be thought of as the treetop or upper part of the tree. If environmental conditions have resulted in a limited amount of food, it is possible that growth regulators will not have initiated cambial activity all the way to the base of the trunk (Figure 1.14). If growth was never initiated on the right side of the lower part of the tree (second and third rings from the outside in Figure 1.14), then rings formed during those years are missing in the lower part. Thus, by virtue of the fact that the tree of Figure 1.14 was alive during those years, it had leaves and had rings in the upper part of the tree, nearer the food source. If an increment core was taken from the right side of the lower part of the stem, it would contain no rings for the second and third from the outside.

Outer rings tend to be narrower than inner rings, and corresponding rings tend to be narrower in the lower than upper parts of the tree (Figure 1.12). Further, growth reduced by environmental conditions in rings that are inclined to be narrower anyway, increases the chances of missing rings. Thus, the

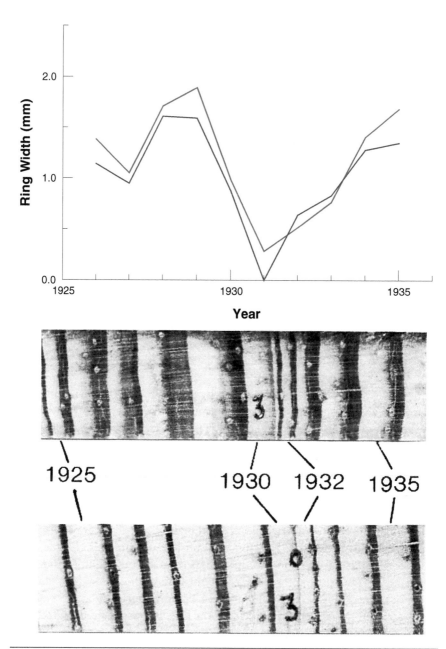

Figure 3.1   Ring-width graphs and photos of segments of two pine cores. The year 1930 is marked as 3 on both cores. The ring representing 1931 is missing from the lower core as designated by the small circle placed on the boundary between the 1930 and 1932 rings. Note the ghost of a 3 in the 1929 ring of the lower sample obviously being misplaced before it was determined that 1931 was missing.

probability that rings will be missing is greatest in the outer rings of the lower trunks in environmentally stressed, older trees and in more slowly growing understory trees.

Discontinuous Rings

Quite commonly, samples from canopy trees in mesic sites contain no missing rings for the years that these trees have been in the canopy. Because of the direct relationship between missing rings and limiting factors, missing rings occur more frequently in semi-arid sites and in understory trees in closed-canopy forests. Also, because trees on sites where environmental factors are not particularly limiting to tree growth tend to produce larger rings, it may be generalized that the larger and more uniform the rings, the less likely are rings to be missing.

A simple analogy can be illustrated by thinking of the hypothetical stem of Figure 3.2 as a pole. Then think of honey poured over the pole as being analogous to new growth. If enough honey is poured over the pole, it will reach the

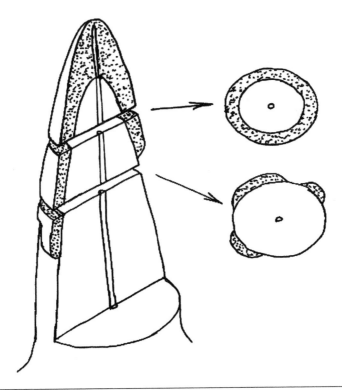

Figure 3.2 Hypothetical stem on which the outside ring (shaded) does not reach to the base. The lower of the two transverse sections taken from the stem shows the outer ring (shaded) as discontinuous.

base. If the honey does not reach to the base, then the result is a missing honey layer (ring) at the base.

Just as a honey layer that does not reach the base may not end at exactly the same height on all sides of the tree, so may a ring (such as the second and third from the outside rings on the right side of Figure 1.14) end at different heights on different sides of the tree. In fact, just as honey may have lobes or *runs*, so may an actual tree ring. If a cross section was taken through these runs, they would appear as a discontinuous ring (Figure 3.2). When dealing with discontinuous rings (Figure 3.3), it is entirely possible that a given ring may appear on one increment core, but not on the core taken from the opposite side of the tree.

For the example shown in Figure 3.3, complete rings were present only in years 1953, 1959, and 1960. Four ring segments (lettered, green, cross-hatched segments, and each composed of one ring) could not be dated from ring-width measurements. Segment 1 was a ring formed in either 1957 or 1958; segment 2

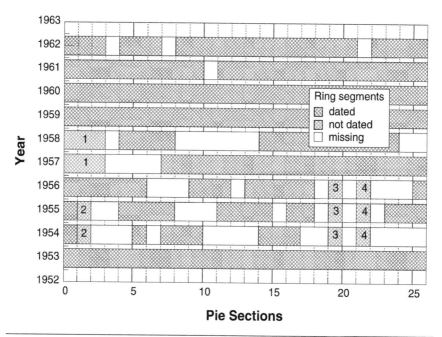

Figure 3.3   Diagram of a series of discontinuous rings from a transverse section of suppressed, sub-canopy red maple (*Acer rubrum* L.), RM-18, from Neotoma Valley research in southeastern Ohio. The transverse section was divided into segments or pie sections, such that each pie section contained no change in ring number; hence, pie section width varied among pie sections.

was a ring formed in either 1954 or 1955; segments 3 and 4 were each formed in 1954, 1955, or 1956. If we divide the cross section into a number of pie sections such that each pie section has a set number of rings, then the width and number of rings can vary between pie sections. Number of rings per pie section ranged from 5 (pie section 4) to 10 (pie sections 15 and 17, the only sections containing all rings during the period of 1953–1962).

False Rings

A false ring superficially resembles a true ring, but is, in fact, contained within a single annual increment, causing that ring to appear as two rings. The obvious difficulty occurs when a false ring is misidentified as a true ring because all tree-ring dates from that point backward in time will be incorrect. False rings are encountered most commonly in conifers and diffuse-porous hardwoods.

During the normal course of a growth season, growth rates decrease, cell diameters become smaller, and cell walls become thicker. This is the typical transition from earlywood to latewood. If, however, during the expected interval of latewood formation environmental factors become less limiting and growth rates increase, then subsequent growth may again resemble that of earlywood. Quite simply, a false ring boundary is formed when greater growth rate resumes after having almost, but not completely, stopped. Breaking of a drought is a common cause of renewed earlywood growth that often results in a false ring. Upon cursory examination, the transition from latewood back to earlywood may appear to form a true ring boundary. The more abrupt the transition back to earlywood, the more the transition appears as a true ring boundary. More than one false ring may occur in a single annual increment, hence, the term, multiple ring.

Close examination shows that at true boundaries—that is, those formed when growth completely ceases—there is an abrupt change in appearance between the last-formed latewood of one year and the first-formed earlywood of the next year. The transition from latewood to subsequent earlywood of a false ring is usually more gradual (Figure 3.4). In some cases, most notably some diffuse-porous species with only subtle differences between earlywood and latewood, it may be very difficult to distinguish between true and false rings. After careful crossdating of ring-width patterns among several samples, it may be concluded that a given ring boundary on a particular sample is false, even though it does not appear so. For this reason, it may be extremely difficult to work with (that is, to crossdate) some diffuse-porous species from certain habitats. For example, rings of gum trees (*Nyssa* spp.) from some habitats are relatively easy to work with, but from other habitats they are nearly impossible.

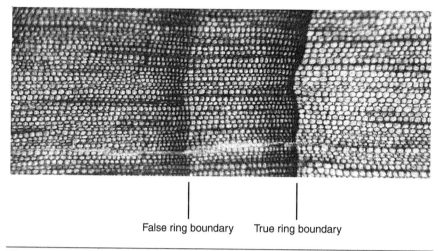

False ring boundary   True ring boundary

Figure 3.4.   A false ring boundary and a true ring boundary.

Similarities in Ring Patterns

Crossdating determines absolute tree-ring dates by matching patterns among samples, or between samples and data from a previously dated tree-ring series. Crossdating would be quite simple if all trees in a given area contained identical patterns of wide and narrow rings, but, of course, they do not. A basic knowledge of factors affecting radial growth may be helpful in understanding why tree-ring patterns exist and why patterns vary among trees. In order to successfully cross-date cores from a collection of trees, there must be some commonalities in the patterns of wide and narrow rings. Similar growth patterns suggest that growth is responding to something in common among trees, namely, regional climate. But, factors must be limiting to growth in order to be correlated with growth. Therefore, within practical constraints, the more limiting those environmental factors are to tree growth, the greater are the chances of similar ring patterns. However, as will be discussed later, this does not mean that only the smallest, most growth-inhibited rings are useful in crossdating. The more easily that the tree-ring patterns may be crossdated, the more likely it is that the tree-ring data may be correlated with environmental factors. Thus, as a general rule of thumb, finding trees that contain good climatic information—*that is, a strong climatic signal*—may amount to little more than finding trees with good year-to-year variation in ring width that permit easy crossdating.

Climatic factors may not always be the most limiting factors to the growth of canopy trees of closed-canopy forest stands in the eastern deciduous forest region. Growth may be limited more by effects of surrounding trees (crowding, shading, and overtopping) than by climatic conditions in common among the trees. Under these circumstances, shade-tolerant species (less sensitive to shading and crowding) may be more easily crossdated than shade-intolerant species. Further, because all trees in the understory are more sensitive to crowding, it may be found that canopy trees are easier to work with than sub-canopy or understory trees. Indeed, we often will work only with canopy trees, and will delete from our analysis the data of rings that were formed while the trees were still in the understory.

A few years ago we sampled some Atlantic white cedar (*Chamaecyparis thyoides* (L.) B.S.P.) just inside the North Carolina state line in the Great Dismal Swamp. Atlantic white cedar was being harvested from the site shortly before annexation of that particular area to the Great Dismal Swamp National Wildlife Refuge. The trees that we sampled were evidently left because they were considered too small to harvest. The ring boundaries of the collected increment cores were readily discernible, but we found them to be not dateable. We could not even crossdate the cores from opposite sides of the same tree, let alone cores among different trees. We concluded that we could not date them because the extreme crowding of these shade-intolerant trees rendered each tree influenced more by its immediate neighbors than by climatic conditions in common among trees. This conclusion was later supported when we sampled some older, less crowded Atlantic white cedar trees from another site in the Dismal Swamp. Those older, less crowded trees showed good ring-to-ring variability and what in the field appeared to be good pattern matches among trees. From this we concluded that they were likely responding to regional climate and should crossdate easily.

When No Measurement Equipment Is Available

Discussions of crossdating methods to follow will also include graphed examples of measured tree rings. It should not be implied from this that actual measurements of ring widths are prerequisite to crossdating. Facilities for surfacing and viewing tree rings without measurement equipment can nevertheless often permit crossdating.

METHODS OF CROSSDATING

The first method that we will discuss is skeleton-plot crossdating, which was developed by Douglass and his associates at the University of Arizona LTRR. It seems likely that the method was perfected before they had sophisticated measurement equipment—which leads us to believe that, obviously, material can be crossdated using the skeleton-plot method even if no measurement equipment is available.

The second method is the extreme-ring match-mismatch crossdating used at the United States Geological Survey Laboratory of Tree-Ring Research (USGS LTRR). Our tree-ring lab was started when Bob Sigafoos enlisted George Smoot to design and build a mechanical stage for tree-ring measurement. Thus, early in the game, we measured first and crossdated later. At that time, data were still entered into a mainframe computer by the use of IBM punch cards. Correcting a single dating error often required considerable re-punching of cards. We quickly realized that it made more sense to crossdate before measurement. Thus, as with the Arizona skeleton-plot method, crossdating at the USGS LTRR could be accomplished with or without measurement equipment. The extreme-ring match-mismatch method of crossdating may be thought of as a simplified variant of the skeleton-plot method.

Crossdating by the Skeleton-Plot Method

Where environmental factors are severely limiting to growth and where missing, false, and discontinuous rings are expected, the skeleton-plot method that was developed by the LTRR at the University of Arizona is a most practical means of crossdating. Skeleton plotting may be performed on surfaced material either before or after ring-width measurement. Thus, it is possible to obtain precise, accurate dating even if there is no intention of ever measuring the rings. As a practical matter, dating before measurement eliminates the necessity to correct the measurement data for missing and false rings.

Proper surfacing of samples and high resolution dissecting scopes are essential for viewing. Very tiny rings, perhaps with only a few cells per radial file, and false rings just cannot be positively identified without proper surfacing and adequate equipment for viewing.

A skeleton plot is a plot of vertical bars in which the length of each bar is inversely related to the width of an individual ring relative to a few rings on either side. A long bar represents a very narrow ring and a short bar represents a wide ring (Figure 3.5). For material from sites in which environmental factors are severely limiting to growth, narrow rings are more important for crossdating than are wide rings. Perhaps this is why people who work with skeleton plots prefer to give small rings the importance of the longest bars. In practice, it may be desirable to distinguish more than just two sizes of rings—wide and narrow. For example, it may make sense to distinguish very wide, wide, average, narrow, and very narrow rings as illustrated in Figure 3.5.

Comparison of skeleton plots for the various samples of a collection should reveal pattern similarities. Close examination may indicate, for example, that some samples contain a tiny ring where others do not, suggesting that a ring is missing in the latter samples.

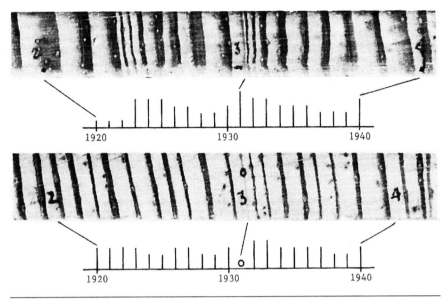

Figure 3.5 Segments of two skeleton plots and corresponding core segments. Note similarities of skeleton plots though one displays a greater rate of radial growth than the other.

Two rings may be represented at a given point on some cores, while only one is represented on others. This might suggest the presence of a false ring in the samples with two rings or a missing ring in the samples without the second ring. If the collection is large enough, crossdating should provide a great probability that all missing and false rings will have been identified. This is based on the assumption that each year is represented by a ring in at least one sample and that each false ring can be positively identified in at least one sample. It is easy to see, though, that as the frequency of missing and false rings increases, so does the probability that something can be missed in the entire collection.

When working with collections containing many missing rings, it may be necessary to not rely exclusively on the collection with which you are working to assure that all years have been correctly identified. The most foolproof way to make sure that all rings are correctly dated is to crossdate the final chronology with a published, dated chronology. The catch is that you may be working in an area for which no published chronologies exist. Some dendrochronologists argue that the next best thing is to have your crossdating checked by another experienced dendrochronologist. Again, though ideal, sometimes this is impossible or impractical. Fortunately, there is still another possibility. Computer programs, such as COFECHA, are available to assist in crossdating. COFECHA will be discussed in more detail later on in this chapter. Generally, these programs do not do the crossdating for you; rather, they are used as a check of the cross-dating that you have done.

All samples in chronologies constructed from living material normally have the same outside date. However, because the sampled trees are likely to be of varying ages, the number of samples containing a given ring will decrease as ring age increases (older, center rings). For example, in a hypothetical collection of 30 samples, all samples might contain rings back to 1700, with fewer and fewer samples containing successively earlier rings, and finally with only one sample containing rings predating 1600. For this example, it might be decided that because only a few samples contain rings prior to 1680, crossdating of rings predating 1680 is unreliable. Therefore, reliability of the crossdating could be qualified with a statement to the effect that the collection was crossdated back to 1680, and that any dates prior to 1680 were based primarily on simple ring-counts wood.

Skeleton plotting is very time consuming but easily justified in situations where missing and false rings are common. More detailed information regarding skeleton plotting can be obtained from *An Introduction to Tree-Ring Dating* by M. A. Stokes and T. L. Smiley and Tree-Ring Research, which is the newer name of the old Tree-Ring Bulletin. There is at least one computer program available that can produce skeleton plots, but it requires measurement data as input, thereby meaning that crossdating with the program before measurement would not be possible.

Crossdating by the Extreme-Ring Match-Mismatch Method

Crossdating, in situations in which missing rings are not common and thus, where environmental factors are not severely limiting to growth, may be accomplished by placing emphasis on rings of extreme size. The extreme-ring method can be thought of as a simplified variation of the Arizona-type crossdating.

An old axiom of plant ecology is that vegetation distribution is controlled by the extremes, not the means, of environmental factors and conditions. This can be restated with regard to ring widths of trees growing where environment is not severely limiting to growth by stating instead that environment has the greatest chances of being limiting to tree growth during years of extreme, not mean, conditions. In arid situations, growth is limited virtually all of the time. In the humid east, there may be years in which growth is not particularly limited at all. Thus, in the humid east, the use of the extreme-ring method employs extremely large as well as extremely small tree rings. This is in contrast to skeleton-plot crossdating in which emphasis is placed on small rings.

The process of crossdating involves identifying extreme-sized rings (both wide and narrow) and counting the number of rings between each pair of consecutive extremes. In practice, a list is composed of the unusual or outstanding

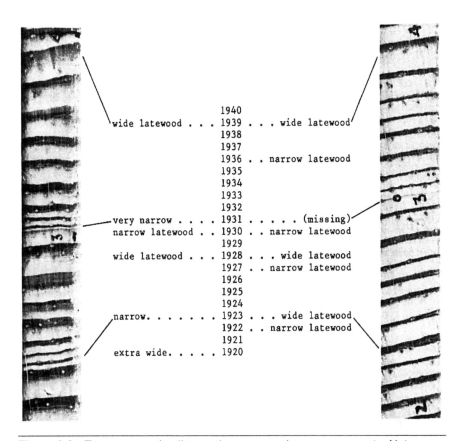

```
                        1940
      wide latewood . . . 1939 . . . wide latewood
                        1938
                        1937
                        1936 . . narrow latewood
                        1935
                        1934
                        1933
                        1932
      very narrow . . . . 1931 . . . . . (missing)
      narrow latewood . . 1930 . . narrow latewood
                        1929
      wide latewood . . . 1928 . . . wide latewood
                        1927 . . narrow latewood
                        1926
                        1925
                        1924
      narrow. . . . . . . 1923 . . . wide latewood
                        1922 . . narrow latewood
                        1921
      extra wide. . . . . 1920
```

Figure 3.6   Two extreme-ring lists and accompanying core segments. Note matches between lists for large as well as small rings.

features of rings of each of the surfaced samples to be dated. In addition to listing unusually wide and narrow rings (Figure 3.6), any other outstanding features, such as false rings and parenchyma bands, are listed. Abnormal features such as frost rings, scarring, and wound tissue are also noted. The product of this effort is a list of distinctive or anomalous years for the collection. This final product can be filed for future reference when dealing with the same, or related, species from similar, nearby habitats.

Determining Which Rings Are Extreme

The designation of *extreme* is rather arbitrary. As with the skeleton-plot method, extreme rings are extreme relative to a few rings on either side of the ring in question. A little bit of practice and experience will indicate whether you are picking too many or too few rings.

More quantitative methods of identifying extreme rings can be used if the ring widths have first been measured. In working with raw ring-width data (that is, in which the non-climatic component or growth trend has not yet been removed), a simple procedure is to describe rings in terms of change from the previous ring. Thus, it makes little difference whether you are working with a series of wide rings or a series of narrow rings. First, calculate the annual sensitivity value for each ring. Then, designate the extremes as those rings for which the annual sensitivity value deviates by more than, say, one standard deviation from the mean of the sensitivity values. The annual sensitivity value, AS, for the nth ring, w_n, is:

$$AS_n = \frac{2(w_n - w_{n-1})}{w_n + w_{n-1}}$$ (Eq. 3.1)

which is really just the change in width from the previous ring scaled to the mean width of the present and previous rings. Annual sensitivity is described in more detail with applications in Chapter 4.

Mean sensitivity, MS, is expressed in terms of absolute change:

$$MS = \frac{1}{n-1} \sum \left| \frac{2(w_n - w_{n-1})}{w_n + w_{n-1}} \right|$$ (Eq. 3.2)

Mean sensitivity is a statistic originated by Douglass and his associates at the Arizona LTRR, though I don't know that they ever used annual sensitivity. MS is often used as a measure of climatic sensitivity and is a very handy statistic that will be referred to in other sections of this book. I suspect that the Arizona LTRR was using it just about as early as when the familiar statistic, standard deviation (σ), was coming into popular usage.

The Extreme-Ring Match-Mismatch

An example of an extreme climatic event to which trees might respond would be a drought. How extreme the growth response is depends on the drought in terms of time and space. For example, all of the trees in a given region may respond to a severe drought. Less severe droughts will be sensed by fewer trees and in more restricted habitats. The geographic coverage of a drought can affect the distance between which collections can be crossdated. As a general rule of thumb, the farther apart two collections are, the more extreme must be the events (such as drought) to which they show time-synchronous responses. In other words, two trees that are close together may have very similar ring patterns, but as the distance between trees increases, similarities in patterns decreases until the only matches between collections are in the most extreme rings.

Individuals of a given species in like habitats may be expected to show similar responses to climate. If either the species or the habitats are different, then the climatic responses may also be different. For example, two different species, such as chestnut oak (*Quercus prinus* L.) and American beech (*Fagus grandifolia* Ehrh.) from the same habitat may respond similarly to some extreme droughts, but may respond differently to the more normal ranges of precipitation and temperature such that their regressions with climate are quite different.

In comparing patterns among trees, even of the same species in a given habitat, do not expect that all the extreme rings of one tree will match all the extreme rings of other trees. Also, expect that many mismatches (opposite responses) will occur when comparing years that are not extreme. For example, a year that is not of particularly extreme size, but also constitutes an increase over the previous year in one tree, may show as a decrease in another tree. The condition that one wishes to avoid in crossdating with the extreme-ring match-mismatch method is a mismatch between extreme rings, that is, an extremely large ring (or extreme increase over the previous ring) of one sample on the same year as an extremely small ring (or extreme decrease over the previous ring) of another sample. In working with graphed data, one can slide a graph of an undated collection along the graph of a dated collection one year, or one step, at a time until the greatest number of matches and fewest mismatches between graphs is achieved.

GENERAL GRANT'S CABIN

The primary purpose of crossdating collections for use in ecological studies is to assure proper dating of events and conditions. For example, did the growth release of the Lynn/Middle Ditch loblolly collection occur on the same year as the growth release of the Hudnel Ditch collection several miles across the Dismal Swamp? If either or both of the collections is incorrectly dated, then we really cannot be sure about the comparison of dates between the two collections. Under ideal conditions for dating, a good, dated collection from a nearby location will be available. Unfortunately, though, this is often not the case. Dating of the General Grant cabin, though not part of an ecological study, is an example of such a situation.

Background

A log cabin was built for General Ulysses S. Grant during the Civil War in the fall of 1864 (Figure 3.7). The cabin was built at City Point in Hopewell, Virginia, as Union troops prepared to lay siege on Richmond. After the war, the residents of the Richmond area would likely have had little concern for a cabin built for

Figure 3.7   General Grant's cabin at City Point, Hopewell, VA. The stockade cabin was built for the general during the Civil War. The end of the cabin overlooking the confluence of the James and Appomattox Rivers may have been originally composed of pine. Tree-ring dates from oak logs of the main part of the cabin (seen here) indicate that the logs are old enough to be original.

General Grant. The cabin was dismantled, moved to Philadelphia, and reassembled in Fairmont Park.

The cabin is a *stockade-type* cabin, meaning that the outer walls are formed by vertically oriented small logs or poles. Henry Magaziner, Regional Historical Architect, National Park Service (NPS) in Philadelphia, told us that a newspaper account written when the cabin was moved from Hopewell to Philadelphia described the cabin as being made of Virginia pine logs. Magaziner further stated that newspaper accounts and other evidence indicate that repairs, including at least replacement of the roof and some of the stockade logs, were made to the cabin at various times in the past.

The NPS wanted to move the cabin back to City Point in Hopewell in 1980. Before going to the considerable expense of actually moving the cabin, NPS wanted to know if any of the existing cabin was original. If none of the original cabin still remained, the historical value would of course have been considerably diminished.

It seemed most likely that the logs used to construct General Grant's cabin in Hopewell were of local origin. Because we had no dated collections from central Virginia, we were not optimistic about successfully crossdating samples from the cabin with samples from any existing, dated collections. In other words, we believed that we would likely have to develop a tree-ring collection from living oak trees in the Richmond area before we could successfully date samples from the cabin.

Reconnaissance

During a reconnaissance trip with Magaziner, we noticed that most of the stockade logs appeared to have no sawed or hewn surfaces. Though bark was not present and the exposed surfaces of the logs on the outside of the cabin were well weathered, many of these surfaces appeared as if they could have been the surfaces to which bark had been attached. A few logs were hewn along the sides that were in contact with adjacent logs. Most of the logs in the walls of the room facing the Schuylkill River in Philadelphia (now facing the confluence of the James and Appomattox Rivers in Hopewell) appeared considerably less weathered than logs in the rest of the cabin. These newer looking logs had sides that had been cut with a circular saw. The outer surfaces had been roughed up and rounded a bit with an ax. Based on their appearance, we considered these logs to be replacement logs.

All the presumed replacement logs were oak (*Quercus* spp.). The logs in the rest of the cabin were weathered enough that, without taking samples from them, identification during the initial trip to the cabin was tentative at best. Some were definitely oak, some appeared as though they could have been pine (*Pinus* spp.), and some appeared to be red cedar (*Juniperus virginiana* L.).

One log that was believed to be oak was sampled to determine whether there were enough rings and enough variability in ring width to produce a datable pattern. Sampling was done with the type of sampler developed at the University of Arizona LTRR and used in archaeological dating in the southwest (Figure 2.3). The sampler cuts a core about 1 cm in diameter and leaves a hole about 1.5 cm in diameter. Laboratory examination indicated that the sample was indeed oak. Ring-width variation and the number of rings appeared to be sufficient to allow crossdating. Thus, at this point, we believed that samples from the cabin probably could be crossdated among themselves, thereby creating what could be referred to as a *floating* chronology. Determining actual dates would then just be a matter of getting an appropriate, dated chronology with which to crossdate and thereby tie down the floating chronology.

Sampling Methods

Following the reconnaissance trip, two collection trips were conducted. A total of 43 samples were taken with a German sampler (Figure 2.4). It cuts a core 2 cm in diameter and leaves a hole 3 cm in diameter. We selected sampling points so that the surface to be sampled displayed minimal erosion from weathering. Sampling was directed toward the center of the log. If the log was sound, we sampled to at least the center. Even though we selected the best looking logs, a number of the sampled logs were not sound.

Sample location was recorded in field notes, and the sample holes were plugged. Prior to sampling, the hole plugs were made from 6 to 7 cm lengths cut from 3.2 cm wooden closet poles. We tapered the plugs slightly with a wood rasp. At the cabin, the plugs were driven in with a hammer until they were flush with the outer surface of the logs. The new wood of the plugs was conspicuous, compared with the weathered (slowly oxidized), almost black exterior of the cabin. Therefore, during our second sampling trip, after installing the plugs, all plugs from both trips were burned lightly (rapidly oxidized) with a propane torch. I do not recall plugging the smaller hole of the initial sample, which had been taken with the Arizona LTRR sampler during our reconnaissance trip.

Crossdating

We crossdated the samples using the extreme-ring match-mismatch method that, as described earlier in this chapter, is a variant of the skeleton-plot dating used at the Arizona LTRR. Each sample was sanded along a side perpendicular to the long axis of the vessels in order to produce a finished, transverse surface. Surfaced cores were examined with a dissecting scope and measured. Examination allowed us to identify the wood and permitted an initial appraisal of datability. Because we found no evidence of missing or multiple rings, we graphically compared the measurement patterns of the cabin cores with dated chronologies. We tried various alignments until we arrived at the one with the greatest number of pattern agreements (matches) and the least number of pattern disagreements (mismatches). In matching samples, more emphasis was given to rings of extreme size. Extreme mismatches (large rings matched with small rings) were avoided if possible.

Wood identification after core surfacing revealed members of both the red oak and white oak species subgenera as well as American chestnut (*Castanea dentata* (Marsh.) Borkh.) and red cedar. None of the logs that we initially thought could have been pine were subsequently identified as pine.

Examination indicated that 16 of the 43 samples removed from the cabin contained too few rings to be useful for crossdating or contained rings that were so deteriorated that measurements would not be reliable. The 29 samples considered for measurement were assigned Roman numeral sample numbers. Two of the 29 samples (XIX and XXVIII) were too rotten to measure. The 29 measured samples included: 4 red cedar, 14 chestnut, and 11 oak samples (Table 3.1).

Table 3.1 Samples taken from General Grant's cabin when it was located in Philadelphia. [Sampled logs were counted from the nearest corner (right, R, or left, L). Walls were arbitrarily numbered clockwise from the wall away from the river. Only samples selected for measurement are listed.]

Wall-Log Number	Sample Number	Species	Notes
2-4R	XII	Red Cedar	
2-6R	XIII	Red Cedar	
3-1R	XXII	Red Cedar	
3-2R	XV	Red Cedar	
2-18R	XIV	Chestnut	
1-7R	VIII, XX	Chestnut	
1-8R	IX, XVIII, XIX	Chestnut	
1-3L	XI	Chestnut	
7-7L	I	Chestnut	
8-19L	XXV	Chestnut	
8-14L	XXIX	Chestnut	
8-5L	XXVI	Chestnut	
8-5L	XXVII	Chestnut	
8-3L	X	Chestnut	
8-1L	VII	Chestnut	
1-3L(W)	XXI	Oak	Left of window
4-5R	XVI	Oak	Replacement log
4-17R	XVII	Oak	Dated
6-6L	V	Oak	Replacement log
6-2L	VI	Oak	Replacement log
7-9L	XXIV	Oak	
7-8L	II	Oak	Dated
7-8L	XXIII	Oak	Dated
7-4L	III	Oak	Dated
7-3L	IV	Oak	
8-17L	XXVIII	Oak	Rotten; deleted

Though the ring boundaries in the red cedar (samples XII, XIII, XV, and XXII) were quite distinct, the samples contained discontinuous rings and there were sharp discrepancies in ring-width patterns among samples. We had no dated red cedar chronology old enough to crossdate with the tree rings that might have been formed in the 1860s. No further work was done with the red cedar samples.

Because the outer surfaces of the suspected replacement logs sampled (V, VI, and XVI) were hewn, an unknown number of outer rings were missing. With the hopes that we could confirm that these samples would represent relatively recent dates, we tried to crossdate them with dated chronologies. We could not get acceptable matches, either among the samples, or between the samples and dated chronologies. Though we concluded that these logs were replacements, we were unable to confirm it using tree rings.

Crossdating the Chestnut Samples

Three chestnut samples (I, XI, and XIV) had too few rings to permit crossdating. During the first half of the 20th century, essentially all American chestnut trees were killed by the chestnut blight (*Cryphonectria parasitica*). Many chestnut trees in central Virginia may have been gone by the mid-1920s, and essentially all were probably gone by the mid-1940s. Though early dendrochronologists may have put together chestnut chronologies, we know of none. Further, as far as we know, no one has yet attempted to construct a chestnut chronology from chestnut material that still exists as building timbers in historical structures.

The American chestnut displays an interesting and unique growth characteristic. Small trees can survive in the understory of a forest for an exceptionally long time. When an opening occurs in the canopy, the trees can bolt into the opening much more quickly than oak trees. This rapid growth in height results in straight trunks that were prized for construction.

American chestnut trees are typically fast growing with wide tree rings. Compared with white oak, similarly sized chestnut logs used in construction generally contain fewer rings and display less ring-to-ring variation in width. Thus, it was not surprising that the chestnut samples we obtained from General Grant's cabin crossdated very poorly with one another. Even samples IX and XVIII which came from the same log, crossdated poorly with each other. Only three of the chestnut samples (VIII, XVIII, and XX) crossdated reasonably well among themselves. We tried to crossdate these three samples with white oak chronologies. White oak and chestnut belong to the same family (*Fagaceae*) and grow in similar habitats. We looked for pattern matches with a white oak chronology from Limberlost in the Shenandoah National Park, and a white oak chronology from eastern Tennessee (collected by Dan Duvick). No matches in patterns were apparent. The chestnuts thus remain undated.

Crossdating the Oak Samples

Of the ten oak samples, reasonable pattern matches were noted among four of the samples (II, III, XVII, and XXIII). Two of the samples (II and XXIII) are from the same log. Unlike the chestnut data, the averaged oak data showed good year-to-year variation, indicating reasonable pattern correspondence among samples.

In the original 1980 report to the NPS, pattern matches were noted between the oak samples of the cabin and the Limberlost chronology—and between the cabin oaks and the Tennessee chronology. During re-examination, we used a chronology from Mountain Lake in southwestern Virginia (collected by Ed Cook) and a chronology from near Knoxville, Tennessee. This Knoxville chronology (A5A) was assembled after 1980 by Sandy McLaughlin's Oak Ridge project, Forest Responses to Anthropogenic Stress. The Knoxville site is about 600 km and the Mountain Lake site is about 200 km from the Hopewell-Richmond area.

The mean of the cabin samples was graphically aligned with the dated Mountain Lake and Knoxville collections, using 1861 as the outside year of the cabin samples (Figure 3.8). This was in accordance with the dating described in the original 1980 report. This alignment showed only one fairly extreme mismatch among collections; that is, between the cabin and Mountain Lake for the year 1834. If the cabin graph is placed in any other alignment, the number of extreme matches is greatly reduced and many mismatches occur. Considering

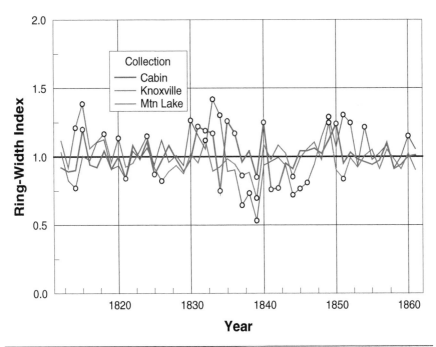

Figure 3.8 Mean ring-width index of four oak samples from General Grant's cabin in central Virginia and a white oak collection from Mountain Lake (Mlake10) in southwestern Virginia and a white oak collection (A5A) from near Knoxville, Tennessee. Index values more extreme than one standard deviation are identified with a symbol (°).

the distance between the cabin and the sites of the dated collections, and realizing that we initially did not expect to be able to date the cabin logs without closer, dated collections from central Virginia, the number of matches and absence of mismatches is impressive. Dating of the cabin logs using data from the Mountain Lake and Knoxville collections confirmed the original dating that had been done in 1980.

The date of the outside ring from the cabin samples determined by crossdating is indicated as 1861. Before averaging data from the oak samples, data from the outside rings was deleted from samples III and XXIII to give all four samples the same outside date for dating purposes. Thus, after dating, the outside date of samples II and XVII remains 1861 and the outside date of samples II and XXIII becomes 1862. Samples II (1861) and XXIII (1862) are from the same log. Weather eroded some outside rings from the log, apparently having eroded one more ring from the sampling point of sample II than from the sampling point of XXIII.

H. J. Heikkenen developed a method of crossdating based on some work by A.W. Ghent. Heikkenen's variation was based on the use of what he called relative ring width (RRW). RRW was either (+) or (−) depending on whether ring width was an increase or a decrease from the previous year. He then used chi square from a 2 × 2 contingency table to test various crossdating alignments. For our purposes, we simply tallied the number of matches in RRW between the 49 years of the cabin samples and the dated chronologies. A match amounted to a RRW of the cabin being of the same sign as the RRW for the same year of a dated collection. This gave the same results as one would get from the sign test, referred to by F. H. Schweingruber as the Gleichläufigkeit score (G-score). The tallied matches in Table 3.2 are expressed as percentages of the 49 rings of the cabin samples. Comparing RRW matches can be regarded as a measure of pattern similarity or dissimilarity. RRW matches above 50 percent suggest pattern similarity and matches below 50 percent suggest pattern dissimilarity. Though this is not a particularly robust test for pattern similarity, the fact that RRW of the cabin oak samples matched the dated collection samples in more than two-thirds of the 49 years suggests pattern similarity.

To determine whether we might be off by one year in our dating, dates of the cabin log samples were moved forward one year and the percentages of

Table 3.2 Percentage of RRW matches (that is, sign matches) among cabin and ring-width collections

	Cabin with Mountain Lake	Cabin with Knoxville	Mt. Lake with Knoxville
RRW Matches	73% (33/49)	71% (34/49)	65% (32/49)

RRW matches out of 48 years were tallied. Then dates of the cabin samples were moved backward one year and the percentages of RRW matches were again tallied (Table 3.3). The RRW matches between the cabin samples and the dated collections are considerably less than 50 percent; hence, the cabin samples are distinctly dissimilar to the dated collections. That the RRW of the cabin samples are quite similar to the dated chronologies for the correct dates and are quite dissimilar when shifted by one year certainly adds support to the 1861 and 1862 dates of the tree-ring dating.

Table 3.3 Percentage of RRW matches when cabin date shifted one year

	Cabin with Mountain Lake	Cabin with Knoxville
Cabin sample dates moved forward 1 year	35% (17/48)	31% (15/48)
Cabin sample dates moved back 1 year	46% (22/48)	40% (19/48)

Conclusions and Final Thoughts on Grant's Cabin

The outside walls of General Grant's cabin consist of logs from two or more oak species, as well as American chestnut and red cedar. We suspect that most of the logs in the three walls of the room that is presumed to be the front room (facing the Schuylkill River in Philadelphia and the confluence of the James and Appomattox Rivers in Hopewell) are replacements. Logs in these walls are less weathered and appear to have been trimmed with a circular saw. Most of the logs in the rest of the cabin are highly weathered and either show no evidence of saw cuts or have sides between logs that were trimmed either by ax or crosscut saw.

Logs in the cabin represent species that are local to the Hopewell area. The Union Army frequently used pioneers to clear wooded areas for rights of way. Grant's troops built a narrow gauge railroad from City Point toward Richmond. It seems possible that the cabin had been constructed using the logs from the local trees that were cut to create the railroad right of way. The tree species of the logs to be used to build temporary cabins was probably of little concern.

Oak and chestnut logs in the walls of the main part of the cabin are badly deteriorated. It seems likely that most of the original logs in the river-facing room were replaced because they were so badly deteriorated. As badly deteriorated as the oak and chestnut logs are, it is not likely that pine logs would have survived until the present. The presumed replacement logs may have replaced pine logs. If the front of the cabin had been pine originally, it is not surprising that newspaper accounts at the time the cabin was moved to Philadelphia would have referred to the cabin as being made of Virginia pine.

Deterioration included considerable rot near the base of the logs as well as extensive damage from insect activity on the interior of the logs. Although the exposed surfaces of most of the highly weathered logs appeared to be bark surface (bark missing, but no evidence of additional material removed by ax or saw), natural weathering likely removed one to several growth rings. I selected sample points where erosion by weathering appeared minimal. From four oak samples, I determined the dates of the outside rings to be 1861 and 1862. If one were to assume no missing outside rings, unlikely as that may have been, it would appear that the trees were cut after the 1862 growth season and before an appreciable amount of the 1863 ring was formed. This would suggest that the cabin was built in the summer or fall of 1863. General Grant's cabin was built in 1864. Because it is uncertain how many rings have eroded away, the 1861 and 1862 dates from tree rings are surprisingly close to the actual build date of the cabin in 1864. If, as speculated, the logs were actually cut in 1864, there would have been fewer outside rings eroded away by weather than we at one point feared there might be. We can say with confidence that, yes, there are logs in the cabin built for General Grant in 1864 that are old enough to be the original logs.

COFECHA, A LITTLE HELP FROM THE COMPUTER

Around the world, several computer programs have been written to assist in tree-ring crossdating. COFECHA, developed at the Arizona LTRR by Richard Holmes, is not intended to replace skeleton-plot crossdating. Rather, it is intended to act as a check on material that has already been dated. Ed Cook, at the tree-ring lab of the Lamont Dougherty Geological Observatory, for example, will not give out data from any collection until it has been checked by COFECHA.

The original copy of COFECHA that we got at the USGS was written for use on a mainframe computer. When John Whiton finally got it running (no small task at that time), we thought we would check it first with one of our collections that we knew had no dating errors. When we ran the program it showed a probable dating error in one of the samples. Though we were confident the sample was all right, John checked it anyway. The sample was long enough that it had been cut in two to be stored in soda straws. The core was cut diagonally as described in the previous chapter. One of the rings that appeared on both core segments was inadvertently measured twice (once from each segment). COFECHA was correct. We were impressed.

The intent of COFECHA is to point out situations in which the dating may be questionable. It is up to the dendrochronologist to go back to the data and to the sample to decide whether or not a dating error exists.

TREE-RING MEASUREMENT

Field Counts and Measurements

Examination of the most recently formed tree rings can be done using standard increment corings, but it can also be done using short cores or plugs removed with an increment hammer. The increment hammer typically permits extraction of a core containing the outer 3–4 or more rings, depending, of course, on bark thickness and ring widths. The use of ring widths to compare growth rates among trees should either be limited to trees of nearly equal diameter, or should take tree diameter into account (see also discussion of ring width and tree size in Chapter 2). One method to account for tree diameter is to simply multiply the width of the segment under consideration (such as a single ring or a five-ring sequence) by the mean diameter of the segment.

Segment width, $r_n - r_{n-x}$, where x = number of included rings, times the mean diameter of the segment, $r_n + r_{n-x}$, is equal to $r_n^2 - r_{n-x}^2$. This figure times pi (π) would be equal to the cross-sectional area of the ring segment being examined, assuming that the ring boundaries describe perfect circles. However, if all you are interested in is comparing segments among trees, you really need not go to the trouble of multiplying by the constant (π).

Taking tree size (trunk diameter) into account provides a better basis for comparison, for example, of current growth of a large and a smaller tree. Examination of only the most recent rings, however, is usually not sufficient to make any judgment on climatic sensitivity or potential for crossdating.

Laboratory Measurement

Aside from using rings to simply determine tree age, most applications require some sort of measurement. I have run across some old studies in which measurement was done with a millimeter rule to the nearest millimeter or with a vernier scale to the nearest 0.1 mm. Depending, of course, on the intended use of the data, examination of growth rates or growth trends in a rapidly growing plantation stand of pines, for example, may not necessitate measurements any closer than to the nearest millimeter.

Most measurement equipment intended for measurement of tree rings is designed to measure to the nearest 0.01 mm. A few machines are accurate to 0.005 mm or even to the nearest 0.001 mm (1 μ). Because of variations in ring width of a single ring, there may be some question as to why bother measuring any more accurately than to perhaps the nearest 0.1 mm. On the other hand, if one were working with Western conifers that contain tiny cell diameters and very narrow rings, width measurements to 0.01 mm may seem too crude. Overall, though, it appears that measurements to the nearest 0.01 mm are adequate for most studies.

Most measurement equipment includes a mechanical stage to hold the material to be measured. When working with measurement accuracies of 0.01 mm, properly surfaced material along with a good quality scope and light source are essential (as described in Chapter 2). Most measurement machines include a dissecting microscope as part of the equipment. An ocular of the scope is usually fitted with some sort of crosshair or reticule. As the mechanical stage moves the material across the field of view of the scope, stage position is recorded whenever a ring boundary intersects the ocular crosshair.

In the early 1970s, we (at the USGS LTRR) used a black and white TV camera attached to the dissecting scope, only to find that we were more heavily dependent on color than we had realized. When we graduated to a color camera, we found that measuring for several hours while watching a TV screen was considerably less fatiguing than hovering over a dissecting scope.

Measuring can be done before the material is crossdated; however, we found it very practical to crossdate first. If you measure first, you will have to edit measured data to accommodate missing or multiple rings discovered in the dating process. We found that when the material was dated before measurement, it was very handy to spot check our location by confirming that we were on the right ring when we crossed a decade mark on the core. When measuring a sample containing hundreds of rings, it is surprisingly easy to skip a ring boundary when momentarily distracted. If not caught at the time of measurement, finding the mistake later can sometimes be time consuming.

Some investigators prefer to work with the transverse surfaces of short stem segments, though perhaps the greatest amount of tree-ring work is done with increment cores. In either case, the axis of measurement travel should be parallel to the nearest wood ray. The intent is that the axis of measurement ends up being perpendicular to the ring boundaries. Because wood rays often do not follow a perfectly straight line, and increment cores may not have been taken parallel to a wood ray, it may be necessary to reposition the sample on the stage one or more times during the course of measuring a single radius.

Years ago, at the USGS LTRR, we measured every radius twice. This allowed us to calculate a number of statistics regarding differences between the two measurement runs and use the mean of the two measurements as the final data of record. The statistics provided several criteria for either accepting or rejecting the measurements. If any statistic indicated rejection, the entire radius (usually from a core) was measured a second time. We learned that after a new operator gained some experience, measurement rejections were quite rare. A practical constraint imposed by double measuring was that the material could not be repositioned during measurement without jeopardizing some of the measurement statistics. When the volume of material to be measured increased (as would be expected with an active laboratory) and when such a small percentage were being rejected, we could no longer justify doubling our measurement time to measure everything twice.

The practice of measuring everything twice provided some measurement statistics that allowed some interesting conclusions regarding accuracy of measurements. At first we were biased in favor of hiring technicians with at least some background in biology, thinking that a better understanding of tree rings would make a difference. Eventually, all of our technicians who were used for measuring tree rings were college students. Beyond that, we couldn't see that it made any difference what their background or training was. They all measured tree rings with about equal accuracy with a minimal amount of training. We did find, though, that most new technicians needed some time to develop proficiency. Some technicians were dismayed that after two or three measurement sessions, their measurement errors were still not as low as they would have liked. Another thing that we found was that for the first hour or so of measurement, males and females measured with equal accuracy. However, though it certainly sounds sexist, the female technicians tended to hold their measurement accuracy longer during a long measurement session than did the males.

Aiming an increment borer toward the geometric center of a tree will result in a core that more or less follows the wood rays, as long as the geometric center is reasonably close to the botanical center. Cores that don't exactly follow along wood rays can raise some measurement decisions. If the axis of the core does not closely follow the wood rays, the usual tendency is for ring boundaries to deviate more and more from perpendicular alignment along the core axis. On the other hand, if growth rates on different sides of the tree have changed over time, then it is quite likely that the wood rays are curved rather than straight. In either case, while measuring the core, one may get to the point of deciding how often to reposition the core in order to maintain perpendicular ring boundaries. What this really boils down to is how far out of alignment the measurement axis can be from the botanical radius before introducing a significant measurement error.

STANDARDIZATION

The Climatic Component

A raw ring-width series may be thought of as year-to-year variation in tree growth, correlative with climate and other environmental factors, which is superimposed on a growth trend (Figure 3.9). Thus, in a generalized way, the year-to-year variation may be regarded as the climatic component and the growth

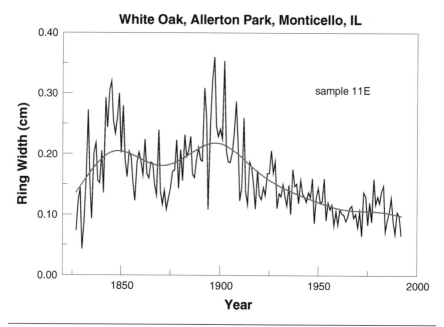

Figure 3.9   Ring-width series of white oak (*Quercus alba* L.). The series may be thought of as being composed of two components: that is, a climatic component (year-to-year variation) superimposed on a non-climatic component (smoothed curve). The data (sample 11E, Allerton Park, Monticello, IL) were smoothed with a 60-year cubic spline.

trend may be regarded as the non-climatic component. The non-climatic trend may be crudely estimated by fitting a curve to the raw ring-width series. Many different methods have been used to fit curves to tree-ring data. A simple curve-fitting function that has come into rather wide acceptance in dendrochronology is the cubic spline smoothing function that was proposed for use with tree rings by Ed Cook and K. Peters.

The climatic component, w_c, can be obtained by simply taking the difference between raw ring width, w_r, and the smooth, non-climatic curve, w_s, derived by fitting a curve to the data:

$$w_c = w_r - w_s \qquad \text{(Eq. 3.3)}$$

The difficulty with the raw climatic component, w_c, is that the amplitude of year-to-year variations decreases with time. In other words, inner rings are larger and are more variable than outer rings. Quite conveniently, the magnitude of year-to-year variability is roughly proportional to ring width from the fitted curve. Thus, the climatic component can be transformed into data that are more time stable by calculating a standardized climatic component, or index, I_c, by scaling the climatic component to the non-climatic component:

$$I_c = \frac{w_c}{w_s}$$

$$= \frac{w_r - w_s}{w_s}$$

$$= \frac{w_r}{w_s} - 1 \qquad\qquad \text{(Eq. 3.4)}$$

The mean of a standardized climatic component series is 0 if the smoothing function has been perfectly applied. The minimum possible value of the standardized climatic component is minus 1.0 and occurs only when a ring is missing ($w_r = 0$). Adding 1.0 to the standardized climatic component results in a standardized index, I, that contains no negative values and a mean of 1.0:

$$I = \frac{w_r}{w_s} \qquad\qquad \text{(Eq. 3.5)}$$

This is the form of the equation to calculate standardized ring-width indices that has been in use since Douglass and his associates established it years ago at the Arizona LTRR. All they were really doing was dividing each raw ring width by its corresponding value on a hand-fitted curve. It is not likely that they thought in terms of climatic and non-climatic components of ring width as we have just described them. In terms of components, the standardized indices may be thought of as an estimate of the standardized climatic component to which a constant of 1.0 has been added. The smoothed ring-width trend—that is, the estimate of the non-climatic component—has sometimes been referred to as the theoretical growth curve. In those terms, we can also think of ring-width index as being actual radial growth relative to theoretical radial growth.

The Non-Climatic Component

Standardization may be thought of as a process to identify and remove the non-climatic component from a raw ring-width series. This leaves the variation in ring width associated with climate (the climatic component). The catch is that there is no strong consensus as to what the non-climatic component should be.

Early work at the Arizona LTRR smoothed the ring width series of a sample by hand fitting a curve to a graphed series of ring widths. They often refer to the fitted curve as the growth trend. The actual process of fitting the curve and removing the trend is often referred to as detrending. At the LTRR they felt, correctly we think, that one could not be expected to do a very good job of curve fitting by hand until having spent a great deal of time looking at and working with numerous collections. When H. C. Fritts joined the LTRR, he tried to describe the process in such a way that it could be calculated on a mainframe computer. Early versions of his program gave the investigator the option of using a straight line (linear regression) or a negative exponential curve. For the high-quality Southwestern collections with which they were working, these options seemed to work very nicely. At that time, Fritts (personal communication) felt that for a few cases the old hand-fitted curves may still have been a better estimate of growth trend, but at least the computer fits could be repeated exactly by anyone.

When Fritts expanded his studies into regions outside the Southwest, he added the polynomial curve-fitting option to his program for standardizing tree-ring data. Later, Ed Cook and K. Peters used a cubic smoothing spline for curve fitting. Their feeling was that in the closed forests of the humid East, the cubic spline did a better job of accounting for common shifts in the growth trend than did the polynomial. In the Southwest, such shifts had either been nonexistent or were often avoided by not processing data from cores displaying shifts.

Working in the Eastern deciduous forest, T. J. Blasing felt that in very broad generalities, the strongest climatic signal was provided using a curve obtained by a cubic spline with a 50% frequency response of about 50 years. This was based on determining the frequency and cutoff length that gave the best climatic reconstruction with *Quercus alba* L. data from Iowa. It has since been generally accepted that the 60-year cubic spline is a good, general-purpose figure to use for most deciduous forest collections. Later, Ed Cook proposed using the cubic spline, again with a 50% frequency response, but with the cutoff in years in the range of $\frac{2}{3}$–$\frac{3}{4}n$, where n is the series length in years. For longer chronologies, this provided a stiffer curve that Cook felt preserved the longer climatic trends (low frequency variation).

As already mentioned, the non-climatic component of classic high-quality chronologies in the Southwest has been described with the negative exponential curve. Richard Holmes concluded that the negative exponential curve did a better job of fitting the early part of the curve of decreasing ring width than did the cubic spline. He further noted that the cubic spline did a better job of fitting the later, flat portion of the curve than did the negative exponential curve. Describing curve fitting of ring-width data as detrending, they proposed double detrending. With this procedure, they first detrended with either a negative exponential or a straight line, and then detrended the result with the cubic-smoothing spline. The stiff curves used in double detrending are intended to preserve long-term climatic trends.

It may be reasoned that if ring-width index (climatic component) is calculated as actual growth (raw ring width) divided by theoretical growth (non-climatic

component), then multiplying modeled theoretical growth (non-climatic component) by real indices could simulate ring growth. Tree rings simulated in this
way were obtained from a forest model. When the simulated tree rings were
treated as real data and standardized, the resulting non-climatic components
were not the same as the ones we started with in the model. Did that mean
that the model was flawed? Because I (RLP) developed the model, I at least
like to think that the model was not completely wrong. Did it mean that accepted methods of standardization are not doing a perfect job of identifying the
non-climatic component from raw ring widths? Well, I should like to think that
this really seems a more likely answer. It would seem that there is plenty of room
for additional work to refine our determination of the non-climatic component.

Stiff Versus Flexible Curves

The separation of a raw ring-width series into climatic and non-climatic components is very much dependent on the flexibility of the curves that are used to
describe the trend in the data. Generally, decreasing the flexibility of the curves
increases the amount of climatic information (particularly long-term or low-fre-
quency information) in the climatic component. Unfortunately, decreasing the
flexibility also increases the chances that non-climatic information will end up
in what we identify as the climatic component. The guessing game is to end up
with a climatic component that contains a minimum amount of non-climatic
information, while at the same time ending up with most of the climatic information. Cook's $\frac{2}{3}n$ rule tips the scale in favor of keeping climatic information,
but does this at the expense of also keeping some non-climatic information in
the climatic component.

In standardizing tree rings for our work with basal area increment (BAI), we
used a 60-year spline. Our intention was that smoothed BAI should represent a
reasonable estimate of the non-climatic component, and we felt that if we used
Cook's $\frac{2}{3}n$ rule, we risked losing some non-climatic information. Further, because we were not interested in the earlier, juvenile rings formed before the trees
attained canopy status, we felt we would not gain anything by double detrending
with the negative exponential.

The bottom line? As we gain more understanding of the standardization process of separating the climatic and non-climatic components, it becomes more
apparent that how we make the separation should depend, at least in part, on
how we wish to use the resulting data. What this may mean in the long run is
that the original ring-width data must be preserved so that individual investigators may standardize with their own *designer* method that has been tailored for
their specific information needs.

Assembling the Mean Chronology

The traditional approach to processing standardized tree-ring data (indices) is to simply merge data from a single species at a single collection site. Merging is most commonly accomplished by averaging the core indices together into a mean collection chronology. It is customarily implied that reference to a chronology, such as *the Kankakee chronology*, really refers to the mean collection chronology.

As early as the late 1960s, Nick Matalas (personal communication), a mathematician at USGS, pointed out that some climatic information held by the tree-ring data of individual trees was no doubt being lost when data from several samples were merged into a mean chronology. He suggested that instead of using multiple linear regressions to describe relationships among several climatic factors (independent variables) and a single mean chronology (dependent variable), it might make more sense to use canonical regression that would allow the use of individual cores. We tried it, but at the time we didn't understand enough of the process to know how to take advantage of what it told us. Later, Hal Fritts used canonical regression to relate regional climatic variables with several mean chronologies.

Historically, averaging is the simplest means of merging data from many samples into a collection chronology. Methods other than the simple arithmetic mean have been used to merge collection data. As an example, one such method is the biweight robust mean. If, in a collection, one sample has an anomalous, large ring for a given year, the value of that outlier will perhaps affect the mean more than you feel is realistic. The biweight robust mean is intended to provide a method to effectively reduce the effect of outliers. Calculation of a simple arithmetic mean probably remains the most popular method of merging collection data and is the method used to produce what is often just referred to as the mean chronology.

Time Series Modeling

A rather sophisticated means of merging collection data has been afforded by time series modeling. Tree-ring data, such as the standardized indices, display a certain amount of serial correlation. This means that the size of any given ring is somewhat related to the size of the preceding ring. Most statistical parameters are based on the assumption that the data are truly independent; that is, that they are not serially correlated. As the serial correlation increases, the strength or significance of various statistical tests and parameters decreases. A correlation coefficient that has not been adjusted for serial correlation may be unrealistically large; that is, the data may not be as closely related as the correlation coefficient suggests.

Time series modeling and pre-whitening techniques with autoregressive moving average (or ARMA) models can provide a means of handling tree-ring data with significant serial correlation. Ed Cook presents a particularly good discussion of standardization and merging into collection chronologies in Section 3: Data Analysis, in the book *Methods of Dendrochronology*, a compilation edited by Ed Cook and L. A. Kairiukstis. Also, a very readable presentation of tree-ring dating and data processing are presented by James H. Speer in *Fundamentals of Tree-Ring Research*.

ELEMENTARY DATA PROCESSING

Numerous computer programs are available that allow calculation of ring-width indices from raw ring-width measurement data. When H. C. Fritts first attempted in the 1960s to standardize the curve-fitting process for calculating tree-ring indices, he wrote a DOS program requiring IBM punch card input to a mainframe computer. This quickly expanded into dozens of programs written by a number of people. Perhaps the two most-used programs are ARSTAN, written by Ed Cook at Lamont-Dougherty, and COFECHA, written by Richard Holmes at the Arizona LTRR. Many of the programs are written in DOS and will run in a data shell on Microsoft® Windows computers. Perhaps more common than the versions to run on Microsoft are versions that were developed for Macintosh computers.

At the time that we at the USGS LTRR began earnestly working with ring area, no tree-ring programs were available that could calculate area. We wrote our own program. This was before we began referring to ring area as basal area increment and so we simply called the program AREA. The original version of AREA, written by Beth Cotter of our lab, was a DOS program written in Fortran. The cubic spline, used for curve fitting to estimate the non-climatic component, was incorporated into the program by John Whiton. Later, Mike Field rewrote the program in program language C, but it still ran as a DOS program. Most computers will no longer run a DOS program without an emulator such as DOSBox. As of this writing, AREA will run with the use of DOSBox and is still available from the authors of this volume.

Because converting ring widths to area is quite simple, you may choose to simply devise your own procedure to calculate BAI. If you wish to calculate either or both smooth BAI and ring-width indices, you will need some type of smoothing function. Remember, any such smoothing function will just provide an estimate of the non-climatic component. Perhaps you can come up with a smoothing function that for your purposes provides a better estimate of the non-climatic component than what you would get from a cubic spline.

SELECTED REFERENCES

Cook, E. R. and K. Peters. (1981). "The smoothing spline: A new approach to standardizing forest interior tree-ring width series for dendroclimatic studies." *Tree-Ring Bulletin* 41: 45–53.

Cook, E. R. and L. A. Kairiukstis (eds.). (1990). *Methods of Dendrochronology: Applications in the Environmental Sciences.* Kluwer Academic Publishers, Dordrecht, Netherlands. p. 393.

Fritts, H. C. (1976). *Tree Rings and Climate.* Academic Press, New York, NY. p. 567.

———. (1991). *Reconstructing Large-Scale Climatic Patterns from Tree-Ring Data.* University of Arizona Press, Tucson, AZ. p. 286.

———. (2001). *Tree Rings and Climate.* Blackburn Press, Caldwell, NJ. p. 567.

Ghent, A. W. (1952). "A technique for determining the year of the outside ring of dead trees." *Forestry Chronicle* 28(4): 85–93.

Heikkenen, H. J. (1984). "Tree-ring patterns: A key-year technique for crossdating." *Journal of Forestry* 82(5): 302–305.

Phipps, R. L. (1979). "Simulation of wetlands forest vegetation dynamics." *Ecological Modelling* 7: 257–288.

Phipps, R. L. and M. L. Field. (1989). "Computer programs to calculate basal area increment from tree rings." U.S. Geological Survey Water-Resources Investigations Report 89–4028, Washington, D.C.

Schweingruber, F. H. (1988). *Tree Rings: Basics and Applications of Dendrochronology.* D. Reidel Publishing, Dordrecht, Netherlands. p. 276.

Speer, J. H. (2010). *Fundamentals of Tree-Ring Research.* University of Arizona Press, Tucson, AZ. p. 333.

Stokes, M. A. and T. L. Smiley. (1968). *An Introduction to Tree-Ring Dating.* University of Chicago Press, Chicago, IL. p. 73.

———. (1996). *An Introduction to Tree-Ring Dating.* University of Arizona Press, Tucson, AZ. p. 73.

4

RING-WIDTH VARIABILITY FOR
ECOLOGICAL INFERENCE

Even when the only purpose is to determine tree age, crossdating is necessary to ensure that the presence of undetected false or missing rings has not resulted in an inaccurate ring count. Measuring the widths of rings, however, provides the starting point for most dendrochronological studies. Even before the rings are measured, most experienced researchers can make a reasonable determination of the degree of ring-to-ring variability of a core by visual inspection. A ring series with little variability is considered *complacent*—that is, when radial growth is fairly uniform from ring to ring despite yearly differences in climate and other factors that are expected to control growth. A *sensitive* series will exhibit considerable year-to-year variations in ring width, suggesting that radial growth is highly correlated with environmental factors. One might expect to encounter highly sensitive trees in the humid East on a rocky hill with thin soil or in a highly xeric environment in the American southwest.

Obviously, then, an objective statistic is necessary to quantify the amount of ring-width variability. Although variability is calculated from the widths of all rings in a core, it may be easiest to illustrate this principle by first simply considering two adjacent rings, as follows:

A hypothetical pair of adjacent rings measure 2.0 mm and 4.0 mm, respectively, with the second ring being the more recently formed. The annual sensitivity, AS, is calculated by subtracting the width of the first ring from that of the second and dividing the result by the average of the two rings: namely, $(4.0 - 2.0) / 3.0$, which results in an AS of $+0.67$. Had the measured ring series been 4.0 mm and 2.0 mm, the AS would be -0.67. Thus, a formula for AS can be written as follows:

$$AS_n = \frac{2\left(w_n - w_{n-1}\right)}{w_n + w_{n-1}}$$

where w_n is the width of the nth ring.

The 2 in the numerator removes the fraction in the denominator that is used to calculate the average of the two rings. An *AS* value can never exceed +2.0 or be less than −2.0. To be convinced that this is true, consider a ring series that would be clearly outside the bounds of reality: for example, 1.0 mm and 100 mm. The *AS* of this series is +1.96. How, then, can a maximum *AS* value range from −2.0 or +2.0? The answer is a series with a missing ring. If a ring of any given width is followed by a missing ring, *AS* is calculated by the width of that ring divided by exactly half its value, thus yielding an *AS* of −2.0 (or +2.0 if the missing ring is the first of the pair). It should be apparent that *AS* can be used to compare ring growth within and among trees regardless of raw ring widths.

The mean sensitivity, *MS*, is the average of the *AS* for each pair of rings that comprise the entire core. As an illustration, consider a core with only four measured rings, oldest to youngest, as follows:

2.0 mm   4.0 mm   2.0 mm   3.0 mm

The respective *AS* values are +0.67, −0.67, and +0.40. The number of *AS* values is always *n*−1 where *n* is the number of rings in the core. The average of the three *AS* values (+0.13) is not a realistic value of ring width variability of the series. Rather, by using the absolute value of each *AS*, an *MS* of 0.58 is generated. *MS* values are always positive but there is no hard rule for a threshold that crosses from complacent to sensitive, although many dendrochronologists consider values exceeding 0.20 to represent sensitive collections.

$$MS = \frac{1}{n-1} \sum \left| \frac{2\left(w_n - w_{n-1}\right)}{w_n + w_{n-1}} \right|$$

ANNUAL SENSITIVITY IN ECOLOGICAL STUDIES

It may seem that *AS* values alone would be of little use in ecological investigations compared to the broader *MS* values that are correlated with growth responses over the entire life of the tree. Indeed, trees with high *MS* values are preferred by researchers who use tree rings to reconstruct yearly climatic variables. However, climate is not the only factor controlling the radial growth of trees. Periodic phenomena such as flooding, landslides, fire, insect outbreaks, and even earthquakes sometimes can have devastating effects on tree growth. For the lack of a more inclusive term, these periodic phenomena will hereafter be referred to as geomorphic events. If the impact of a given event results in a change in the width or anatomy of a ring or series of rings, a number of questions can potentially be addressed. Is the response of impacted trees sufficiently different, and thus quantifiable, compared to corresponding rings of nearby trees unaffected by the event? Can an event that occurred prior to the keeping of instrumented records be documented from the tree-ring record? How often have similar events occurred in the past? Can smaller, more frequent events be identified from larger, less frequent events by the type of tree-growth response? If the aim is to establish a time series of events over perhaps hundreds of years, the first step is to understand how events in the present impact the radial growth of trees. Only then can the present be used to understand the past.

In the following pages, we will discuss the use of both *AS* and *MS* values to document the frequency and magnitude of geomorphic events. *AS* values can be used to determine the immediate effects of any of these periodic factors that are expected to have an impact on growth. We will first examine growth responses to an event occurring during the period of dormancy and will then consider an event that occurred during the interval of active radial growth. In each case we have chosen an event of extremely high magnitude because it seems reasonable to expect that radial growth responses likewise might be extreme.

Events Occurring During Dormancy

An event occurring during dormancy is expected to affect growth only in the subsequent growth year. Because the first ring in the *AS* pair formed prior to the event, the *AS* value will depend in large part on the width of the first ring in the series. In other words, if the first ring is unusually wide or unusually narrow, the *AS* value may not adequately reflect the impact of the event. For example, if an event severely damages a tree so that subsequent growth is minimal, the *AS* value will overestimate the impact of the event if the first ring is unusually wide; the *AS* value will underestimate impact if the first ring is unusually narrow. To better evaluate growth responses, the *AS* values of impacted trees should be compared to nearby trees of the same species unaffected by the event in question. Roughly speaking, this minimizes the effects of climate. This may be difficult if the event impacts large areas, such as insect outbreaks or wildfires. More localized events such as slope failure or flooding may afford the opportunity to locate nearby trees outside the impacted area.

The United States Geological Survey (USGS) Laboratory of Tree-Ring Research (LTRR) has conducted studies of flood-plain trees since the 1950s in order to determine the relationships between flooding and tree growth. Several studies were conducted along two reaches of the Potomac River upstream from Washington, D.C. One site is near Chain Bridge where flow velocities during even minor floods typically exceed 6.7 meters per second (m/s). A broad bedrock floodplain supports woody vegetation that is kept small and shrubby from frequent pruning by debris-laden flood flows. A second reach is near the mouth of Difficult Run about 13 km farther upstream. Unlike the straight channel at

Chain Bridge, the river here flows along a series of channel bends and, as a result, flood-flow velocities are greater along the outside than the inside bends. Woody vegetation ranges in form from small and shrubby where flood flows are greatest, to stands of tall, full-canopied trees growing on thick alluvium along reaches protected from damaging flow velocities.

A small, ice-laden flood (peak flow about 2,100 m³/s) in February 1948 resulted in ice jamming at Chain Bridge that covered parts of the bedrock floodplain to depths of as much as 12 m. At present, the oldest trees along most of the floodplain have a 1948 center ring, having grown as sprouts from roots anchored in the bedrock, but not killed. However, Robert Sigafoos, a botanist and founder of the botanical studies unit of the USGS LTRR, was able in 1960 to recover samples from six ash trees (*Fraxinus*) that had been growing prior to 1948. Unfortunately, notes pertaining to their precise locations and growth forms were not retained. We were able to locate five ash trees growing at higher elevations believed to have been unaffected by ice jamming.

AS values for the 1947–48 ring series ranged from −0.07 to −0.58 ($\mu = -0.33$) for the six samples collected by Sigafoos. The five unimpacted ash trees had *AS* values ranging from +0.4 to +0.27 ($\mu = +0.16$). The *AS* values of the latter group seem reasonable considering that precipitation during the 1948 growing season was greater than that during 1947 and thus would be expected to favor greater radial growth.

No information is available concerning ice jamming in 1948 at the Difficult Run site. However, considerable jamming that occurred at Chain Bridge in January 1968 (though on a much smaller scale than in 1948) was not observed along the Difficult Run reach. It seems likely however, that flood flows during February 1948 were ice laden and would be expected to have damaged vegetation. Five ash trees that grew along reaches where flood-flow velocities were greatest (2 to 3 m/s) had *AS* values for 1947–48 ranging from −0.26 to −0.70 ($\mu = -0.45$), whereas 10 ash trees growing above the crest elevation of the flood showed *AS* values ranging from −0.03 to +0.69 ($\mu = +0.19$). This average was comparable to that of unimpacted trees at Chain Bridge.

The intent of this exercise was simply for demonstration purposes. Any hard-and-fast conclusions are tenuous because no actual observations were made of the trees prior to or just after the 1948 event. Field studies are essential to gather information concerning the precise location of study trees, their size and form, and any additional information that may be helpful once increment cores are studied in the laboratory. The concept of *random sampling* is not pertinent to most tree ring studies. If the intent is to recover evidence of flooding, trees should be studied that have downstream-leaning trunks that bear abrasion scars or other signs of flood damage. On the other hand, a researcher interested in reconstructing drought histories would most likely find little success studying flood-plain vegetation.

Events Occurring During Active Radial Growth

A geomorphic event occurring during the growing season would be expected to possibly impact growth for the remainder of that year as well as that of the following year or years. As an example, the USGS studied woody flood-plain vegetation following a flood on the Potomac River in 1972. The flood crested on June 24 with a peak discharge ($10{,}170$ m^3/s)—exceeded during the previous 100 years only in 1889, 1936, and 1942. Unlike the 1948 ice jam, vegetation along the Difficult Run reach was studied prior to the flood and detailed records of trees were made afterward.

Ash trees varied greatly in size, form, and age prior to the 1972 flood. The 1 km long study reach supported small, shrubby trees like those at Chain Bridge, but also forests of larger trees along the inside of channel bends and just downstream from protective outcrops that sheltered trees from damaging flood flows. The 1972 flood overtopped many of these protective features and, as a result, entire swaths of trees were uprooted or severely damaged. Many surviving trees were partly uprooted and leaning downstream, and many had damaged or missing crowns. In the zones where damage was greatest, most surviving trees had most or all of their leaves stripped away. Soon after the flood, many trees produced a new crop of leaves. As will be discussed in Chapter 5, some trees (such as ash) that produced a second crop of leaves following the flood also formed a band of large vessels within the 1972 latewood. The amount of radial growth after the band of vessels was often minimal in trees with severely damaged crowns and often greater in trees showing fewer external signs of damage. Thus, as might be expected, radial growth for the remainder of 1972 seemed to depend on the amount of flood damage and perhaps on the capacity for recovery.

The AS values for the 1972–73 ring series were highly negative for some trees and highly positive for others. Cores were taken from 23 ash trees growing on surfaces flooded on average twice yearly. Sixteen had crowns that were inundated by the 1972 flood. The AS values seemed to be correlated with the degree of crown damage. Six trees with heavily damaged crowns had AS values for the 1972–73 ring series that ranged from -0.62 to -1.59 ($\mu = -1.00$); seven trees with less severe damage had AS values ranging from -0.01 to -0.37 ($\mu = -0.24$). Three trees, however, had AS values of $+0.51$, $+0.78$, and $+0.78$, respectively ($\mu = +0.69$). These trees were defoliated by floodwaters but their crowns were intact. However, they grew previously amidst a group of trees uprooted and washed away during the 1972 flood, and thus their wide 1973 rings may have resulted from a growth release like that observed among trees left standing after neighboring trees are harvested. Trees growing at Chain Bridge on surfaces flooded on average from 1.25 to 2.0 years also showed a wide range of AS values (-1.47 to $+1.24$) depending on the degree of crown damage. Interestingly, none of the trees growing at the lowest elevations of the bedrock floodplain showed extreme positive or negative AS values. Trees there typically do not exceed 2 m in height and have sparse, flood-trained crowns. Field observations suggest that damage is most likely from flood-borne debris when the river stages coincide with the height of trees, as occurs on average about three times yearly. Once a tree is totally submerged, however, even great floods may be unlikely to severely damage their crowns or dislodge roots that are firmly anchored in the bedrock.

The study of flood-plain trees by the USGS was initiated in the 1950s when flow records for most rivers and streams were either unavailable or extended only to 1930. Thus, the aim of studies was to develop techniques that could identify floods that occurred before the period of record, thus providing a more realistic estimate of the recurrence intervals of floods. Regardless of hydrologic implications, the radial growth of flood-plain trees can be used to investigate numerous ecological questions. The growth of flood-plain trees is controlled in part by climate, but raises the question: Are climatic factors more or less important than flooding? Are these relationships consistent among different species— or even among a single species subjected to differences in the frequency and magnitude of flooding? Does the capacity to recover from severe flood damage determine the species compositions of flood-plain forests? Under what circumstances are trees of the same or different species expected to crossdate?

MS IN ECOLOGICAL STUDIES

Unlike AS, which measures the change in width from only one ring to another, MS provides a measure of ring-width variability during the entire life of a tree and can be used to compare trees regardless of whether they are fast or slow growing. Furthermore, if desired, MS values can be compared for selected groups of rings in common to all trees in a collection—for example, comparing the MS for rings that formed from 1940 to 1960 for all trees in a collection. If a persistent factor, such as perhaps regional air pollution, was suspected to begin impacting growth in, say, the early 1960s, calculating the MS from that year forward might provide evidence of a commensurate change in the growth of trees. Similarly, a single average MS can be calculated for an entire collection and compared to similar chronologies where factors controlling growth may be different.

MS values of flood-plain trees at the Chain Bridge and Difficult Run sites were used to determine if radial growth is controlled primarily by flood frequency (that is, how often trees are flooded) or by flood magnitude (the discharge of peak flows). This requires field observations to match the discharge of a given flow (determined from a rating table from a nearby gauge) to the various elevations where study trees grow. We also determined the approximate velocity during various flood flows from bank observations and from a canoe. Thus, for each study tract we had data for flood frequency and current velocities—the latter considered a proxy for the potential for damage to trees.

A group of 33 ash trees selected at the Difficult Run site grew on surfaces that flooded approximately twice a year. The smallest floods covered the bases of trees to a depth of 1 m, but to as many as 5 m during floods that occur on average every two years. Although trees were all flooded by the same approximate discharge, they were divided into three groups based on estimated flow velocities: a *full-exposure zone* where current velocities were 2 to 3 m/s, a *zone of sheltering* where velocities were less than 1 m; and an *intermediate zone*. Current velocities within the three zones differed due to the hydraulic geometry of the channel along the 1 km study reach.

The *MS* of trees was greatest in trees growing along unsheltered reaches of the channel, ranging from 0.36 to 0.49, whereas those in the sheltered zone had values ranging from 0.17 to 0.29. The *MS* of trees in the intermediate zone (0.30 to 0.38) barely overlapped the *MS* of the other two zones. Furthermore, trees sampled from the sheltered reach but growing on surfaces flooded from one to every two years had *MS* values nearly identical to the sheltered trees growing on frequently flooded surfaces. Thus, at least at this site, *MS* is not correlated with flood frequency. Rather, the intensity of flooding, in turn a function of channel geometry, appears to control much of the form, age, and radial growth of these trees.

A second example of the utility of *MS* values is a study of saline encroachment in a forested wetland in the Blackwater National Wildlife Refuge (BNWR) in Cambridge, Maryland. Most of the BNWR is less than 1 m above sea level, and aerial photography taken since the 1930s shows an expansion of open water within much of its central area. We collected cores and cross sections from 34 loblolly pines (*Pinus taeda* L.) growing within a 30 by 110 m plot along a suspected gradient of saltwater intrusion. Six trees were already dead at the onset of the study. The forest is even-aged and began to grow in approximately 1940. We used a laser level to establish surface elevations above mean sea level for each tree to the nearest 0.5 cm from the average of the closest four to six elevational measurements within the grid. The elevations of trees ranged from approximately 82.0 to 34.5 cm above mean sea level. Table 4.1 lists trees in order of their elevation above mean sea level.

MS values of the 34 study trees were strongly correlated with growing positions relative to mean sea level (Figure 4.1). Ten trees growing at elevations higher than 55 cm had *MS* values that ranged from 0.17 to 0.31, but only two trees had values equaling or exceeding 0.30. The *MS* values of six trees growing at elevations between 53.5 and 43.5 cm ranged from 0.24 to 0.34, but four trees had values that equaled or exceeded 0.32. However, the remaining 18 study trees

Table 4.1 Summary of elevation and growth data for 34 loblolly pines from the Blackwater National Wildlife Refuge, Cambridge, Maryland. [Mean sensitivity is a measure of ring-to-ring variability of each tree, and thus an estimate of responses to environmental factors controlling radial growth.]

Tree	Elevation Above Mean Sea Level (cm)	Diameter at Breast Height (cm)	Years Measured	Mean Sensitivity	Status April 2007
1	82.0	38.2	1943–2005	0.22	Living
2	81.0	31.4	1944–2005	0.21	Living
3	75.0	30.6	1942–2005	0.27	Living
4	74.5	30.7	1943–2005	0.24	Living
5	73.5	23.6	1942–2005	0.31	Living
6	72.4	33.8	1940–2005	0.30	Living
7	69.5	34.3	1943–2005	0.25	Living
8	62.5	46.7	1940–2005	0.23	Living
9	58.0	39.6	1942–2005	0.25	Living
10	55.5	34.6	1943–2005	0.17	Living
11	53.5	30.2	1940–2005	0.32	Living
12	50.5	23.2	1943–2005	0.32	Living
13	48.0	30.4	1942–2005	0.24	Living
14	44.0	20.2	1944–2005	0.29	Living
15	43.5	19.0	1944–2005	0.34	Living
16	43.5	20.2	1942–2005	0.33	Living
17	40.5	16.7	1940–1996	0.34	Dead
18	40.0	12.6	1942–1996	0.39	Dead
19	40.0	26.3	1940–2002	0.38	Dead
20	39.5	14.8	1940–2002	0.48	Dead
21	39.5	8.9	1951–2001	0.56	Dead
22	39.0	15.9	1940–1997	0.35	Dead
23	39.0	13.3	1940–1996	0.49	Dead
24	39.0	16.9	1941–2002	0.50	Dead
25	38.5	18.0	1940–2002	0.48	Dead
26	38.0	10.9	1940–1994	0.48	Dead
27	38.0	8.9	1945–2002	0.56	Dead
28	38.0	10.7	1941–1995	0.43	Dead
29	37.5	21.0	1941–2002	0.49	Dead
30	37.5	12.7	1940–2002	0.72	Dead
31	36.5	17.7	1940–2002	0.57	Living
32	36.5	15.8	1940–2001	0.59	Dead
33	36.0	15.9	1940–2002	0.77	Dead
34	34.5	19.2	1942–2002	0.77	Dead

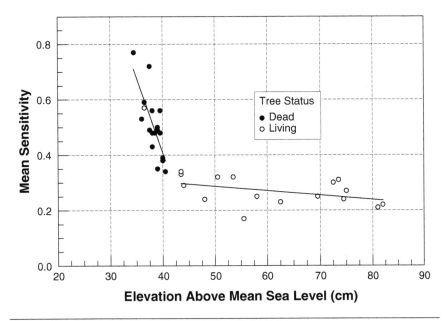

Figure 4.1   Mean sensitivity values of 34 loblolly pines growing along an elevational gradient, BNWR. The elevation at each tree is an average of four to six measurements taken with a laser level. *MS* is more highly correlated with elevation (R^2 = +0.83) than with diameter at breast height (R^2 = +0.59).

grew at elevations ranging from 40.5 to 34.5 cm, and *MS* values ranged from 0.34 to 0.77. Of the ten trees growing at the lowest elevations (less than 39.0 cm), none had an *MS* value less than 0.43. Of these, three were dead at the onset of the study, and six others died within about two years.

The striking differences in *MS* values of trees along an elevational gradient along with the death of most trees that became established on the lowest surfaces strongly suggests that a factor other than climate (hypothesized to be saline encroachment) controls tree growth and mortality at this site. Furthermore, the *MS* values of trees growing at elevations greater than approximately 55 cm above mean sea level suggest that salinity is not impacting their growth. If so, one would expect that radial growth of these trees would show a trend of increasing ring-area growth (basal area increment, see Chapter 8)—and indeed, this was observed. Some trees had area growth exceeding 20 cm² in their outermost rings. Trees that died between 1994 and 2002 typically had reduced ring-area growth (often as little as 1.0 cm²) for two to seven years.

If the hypothesis regarding saline encroachment is correct, soil salinity would be expected to be greater at lower than higher elevations within the study site. Figure 4.2 shows an increasing concentration of sodium in soils collected

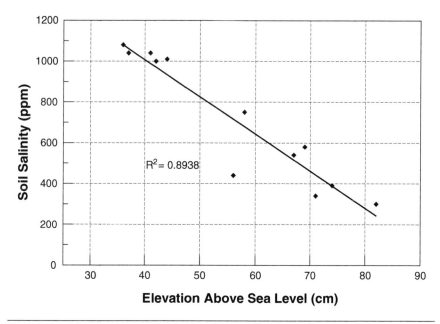

Figure 4.2   Concentrations of sodium (parts per million) in soil samples taken along an elevational gradient, BNWP. R^2 = +0.89.

along an elevational gradient. The smallest and largest sodium concentrations were measured at the highest and lowest elevations, respectively. It seems that loblolly pines are able to tolerate sodium concentrations of less than about 600 ppm (parts per million), but not those in the range of 1,000 ppm. Considering that the relation between soil salinity and elevation seems comparable to that between *MS* and elevation, would it be possible with further study to establish the utility of simply using *MS* values as a proxy for the much more expensive testing of soil salinity?

Trees growing at elevations ranging from 53.5 to 43.5 cm above mean sea level may be at risk from high concentrations of sodium in soils. Trees 15 and 16 (see Table 4.1) both grow where soil salinity is approximately 1,000 ppm, but it is unknown how long these levels have persisted. The MS values for trees 15 and 16 were 0.28 and 0.24, respectively, for the years 1960 through 1980. These values are comparable to those of trees growing on the highest elevations, suggesting that trees 15 and 16 were not subjected to damaging levels of soil salinity before 1980. However, *MS* values of the two trees from 1981 until the present were 0.40 and 0.62, respectively. *MS* values of trees growing at the highest elevations did not differ significantly during 1960–1980 compared to 1981 to the present. Similarly, the *MS* values during both time intervals were 0.20 and

0.28 for tree 14, which grows at an elevation of 44.0 cm. On the basis of *MS* values alone, it seems likely that trees 14, 15, and 16, and possibly some trees at even higher elevations, will die as the area continues to convert to marshland.

SELECTED REFERENCES

Sigafoos, R. S. (1964). "Botanical evidence of floods and flood-plain deposition." U.S. Geological Survey Professional Paper 485A. p. 35.

Yanosky, T. M. (1982). "Hydrologic inferences from ring widths of flood-damaged trees, Potomac River, Maryland." *Environmental Geology* 4: pp. 43–52.

———. (1982). "Effects of flooding upon woody vegetation along parts of the Potomac River floodplain." U.S. Geological Survey Professional Paper 1206. p. 21.

5

PRACTICAL APPLICATIONS

Contemplating the use of tree rings to reconstruct environmental variables such as monthly or yearly precipitation, temperature, or streamflow can seem a daunting task to a novice dendrochronologist. Since the first reconstructions in the 1970s, methodologies have become increasingly mathematically refined and have come to demand commensurate computer skills, as well. However, numerous relatively simple techniques often permit the researcher to draw important inferences concerning both historical and present-day ecological conditions. These methods are based on an understanding of how trees grow and how they respond to stresses that result in changes in growth habit, radial growth, or ring anatomy. This chapter briefly discusses these *simple* kinds of botanical techniques along with a brief explanation of the biological basis for each, as well as examples and suggestions for their use. Although discussed separately, most researchers use several techniques together when approaching a problem in the field.

TREE AGE

Natural disasters such as floods, glaciation, slope failure, and fire sometimes destroy large swaths of forest. New surfaces that were subsequently colonized by woody vegetation eventually support mixed-age forests where the oldest trees are all the same age. If the number of years since the event is within the lifespan of colonizing trees, the synchronized ages of the oldest trees, plus an estimate of the time required for trees to become established, provide a good estimate of the approximate year the event occurred. In other words, if the maximum lifespan of colonizing trees is about 400 years and the oldest trees are all 250 years old, it then seems reasonable to conclude that the event that destroyed the original forest occurred at least 250 years ago. The event occurred more than 400 years ago if the forest lacks an age-synchronous cohort of oldest trees, and thus 400 years is a minimum estimate for the age of the event. This technique is most useful in areas where trees have the greatest lifespans.

An example of this technique is the study of debris flows by the United States Geological Survey (USGS) along several glacially-fed streams on Mt. Shasta, California. Outburst flooding along the steep channels carries sediment-filled and debris-laden flows that destroy everything in their path. Flood deposits are easily identified and it is estimated that surfaces become sufficiently stabilized within 5 to 10 years to support the growth of trees. The USGS used the synchronized ages of trees to document numerous debris flows occurring back to about 1580. Similarly, the lateral and terminal moraines of glaciers become vegetated after the glacier begins to recede. The terminal moraine reveals the maximum extent of glacial advance, and the age of the vegetation that it supports provides a good estimate of the onset of glacial retreat.

Variations on the use of synchronized tree ages also can provide evidence for the occurrence of historical events and processes. For example, the rate of sedimentation in wetlands and along many streams can be determined by placing clay feldspar pads and subsequently measuring deposition over one to several years, but this may not accurately reflect the rate of sedimentation during previous years. The USGS studied forested wetlands where prolonged sedimentation resulted in trees having a *telephone pole* habit rather than the flaring of trunks typical of trees in nondepositional environments. A steel rod was used to locate the buried root collar, which was just below the original surface when the tree germinated. An annual rate of sediment deposition near each tree was determined by dividing the depth of sediments covering the root collar by tree age. An average determined from numerous trees seems to provide a good historical estimate of deposition rates.

Dead trees, preferably those that remain standing, provide the means to determine the year of death by crossdating with nearby living trees, as was discussed in Chapter 4 for loblolly pines growing in the Blackwater National Wildlife Preserve in Maryland. Although this technique is typically used to determine which recent factor or factors negatively impacted tree growth or resulted in death, it sometimes can be applied to trees that died decades or even hundreds of years ago. A tree killed by the Cascadia earthquake that occurred in 1699 or 1700 in Washington and northern Oregon was used to date the event. Fragments of dead trees buried by a large avalanche in Yosemite, California, were crossdated with nearby living trees to determine that the event occurred after the cessation of growth in 1739 but before growth resumed in the spring of 1740. Although these examples require the date of the outermost ring, they nevertheless can be used to date a geomorphic event in a way that is analogous to that of using the date of the innermost ring in the studies mentioned previously.

CHANGES IN GROWTH HABIT

A tree growing in the open on thick, rich soils typically differs in appearance from one growing on a thin, rocky slope crowded with neighboring trees. Trees established along high-gradient streams are often small and shrubby and have small, asymmetric crowns. Many lean downstream as the result of being partially uprooted from frequent flooding (Figure 5.1). Trees growing in open, windswept locations often produce branches only along the downwind side of their trunk, and trees growing on hillsides often lean downslope following episodes of slope instability.

Figure 5.1   Ash (*Fraxinus* spp.) growing along the Potomac River in Maryland, on a surface flooded numerous times yearly. Note the downstream-leaning trunk bearing numerous impact scars and a depauperate crown. Photo by T. M. Yanosky.

The partial uprooting of some tree species stimulates the production of adventitious sprouts along the leaning trunk. Tilting during the growing season often results in sprouting within a matter of weeks; sprouting is likely the following growing season if tilting occurs during dormancy. Sometimes multiple sprouts form but over time, typically only one persists. However, subsequent damage to the tree can result in additional sprouting, and thus a single tree can support sprouts of different ages, each of which is a response to a separate environmental perturbation. Occasionally we have encountered a tree with a split-trunk growth form that developed following shearing of the original trunk without also tipping it. Coring the base of a sprout to determine its age can be used to document the episodic history of the event or events that resulted in the production of sprouts.

Robert Sigafoos, a USGS botanist, began studies in the 1950s along the lower Potomac River near Washington, D.C. in order to determine if the age and form of riparian trees could be used to document the occurrence of historical floods. If so, he reasoned, these techniques could be applied to streams where flow records were incomplete or lacking, thus improving estimates of the recurrence interval of floods. Daily flow records for the lower Potomac River were continuous since 1930 and since 1895 at Point of Rocks, Maryland (67 km farther upstream), thus providing established flow data that Sigafoos could compare against flood histories inferred from flood-plain trees. His classic 1964 paper established the utility of using botanical evidence of floods, and to this day it is not unusual to see a reference to leaning, sprout-origin trees as *Sigafoos trees* (Figure 5.2). To give but a single example, Sigafoos located an ash tree (*Fraxinus* spp.) in which the entire crown was destroyed by a catastrophic flood in October 1942. The tree produced a single sprout in 1943 that eventually replaced the old crown. An ice jam in February 1948 destroyed the new crown and resulted in a sprout that grew from remnants of the 1943 sprout. The tree was damaged again by a large flood in 1955 and subsequently formed a series of new sprouts. Thus, a single tree preserved evidence of three separate hydrologic events.

In addition to often producing sprouts along a tilted trunk, radial growth generally changes along the axis of lean-in response to gravitational stresses. This reaction wood results in eccentric ring growth termed *compression wood* in coniferous species and *tension wood* in hardwoods. Compression wood produces wider rings on the downslope than on the upslope side of the trunk, whereas tension wood is comprised of wider rings along the upslope axis of lean. In either case, the abrupt shift from reasonably concentric to eccentric growth can be used to establish the approximate year that the tree began to lean. For example, the USGS investigated landslides along an incised alluvial fan in the Blue Ridge Mountains of central Virginia. The presence of reaction wood

Figure 5.2 Sprouts of a white oak (*Quercus alba* L.) growing from a tilted trunk on a bedrock terrace along the Potomac River near Washington, D.C. The sprout on the right (diameter 21.6 cm) started to grow in 1937 following a great flood in April of that year. The sprout on the left could not be dated, but most likely also began to grow in 1937. The bag is a reference for size. Photo by T. M. Yanosky.

was used to document slope instability in 1937, 1972, 1993, 1997, and 1999. Each year is associated with local flooding events in nearby streams. Figure 5.3 shows ring widths of two radii along the axis of the lean in the sprout of a white oak (*Quercus alba* L.) that was rotated and tipped in the early 1970s during an episode of slope instability that was most likely caused by intense rainfall in the wake of Hurricane Agnes in June 1972.

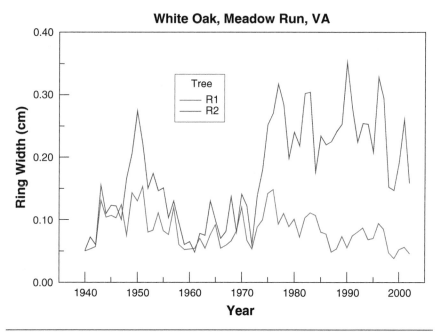

Figure 5.3   Raw ring widths of a white oak (*Quercus alba* L.) sprout growing on an alluvial fan along Meadow Run, Virginia. Wider rings (tension wood) formed along the upslope radius after the sprout was rotated and tilted as the result of slope instability presumably caused by heavy rainfall in June 1972 during Hurricane Agnes.

IMPACT AND ABRASION SCARS

Instantaneous damage to the cambium by landslide debris, fire, or flood-borne ice and debris typically results in the scarring of trunks and branches. Scars provide both a means to determine the year that damage occurred and, in the case of flood scars, to estimate the minimum stage of the flood. In response to localized cambial damage, trees produce an undifferentiated, often discolored tissue termed *callus*. Although a scar may remain visible for many years, eventually it is subsumed by new radial growth and thus may be detected only by examining cross sections of the trunk. It is not unusual for individual trees to bear scars from numerous different events, and we have observed trees with such extensive scarring along their upstream-facing trunks that it is not possible to delineate individual scars.

I (TMY) once gave a talk to a local high school class about the ways that tree rings can be used to learn about historical flooding. When I mentioned that scars can be used to estimate the height of floodwaters, a puzzled student remarked that this seemed unlikely because over time the scar would grow in height as the tree did. I explained that trees grow in height only from their tips and that the scar therefore remains at its original height. However, if the tree is subsequently tipped by another flood, care must be taken to determine that the scar formed prior to the tipping. A reasonable way to do this is to compare the year of scar formation to the year that the tilted trunk began to form reaction wood.

Unfortunately, determining the year of scar formation is sometimes difficult if not impossible, particularly if a scar, in turn, is damaged by later floods. One technique is to take a core sample directly through the face of the scar to the centermost ring; a second core is then taken from the opposite side of the trunk and the difference in the ring counts is the number of years since scarring occurred. However, this method becomes problematic if damage to the bark and cambium extended to underlying rings. Many researchers find a more accurate method is to use a small saw to take a wedge-shaped sample along the margin of the scar that permits counting the number of rings formed since the cambium was damaged.

The scars discussed so far are sometimes referred to as impact scars. Abrasion or rubbing scars form when an object such as a floating log is lodged against a tree. As the water recedes, the log slowly abrades the cambium and leaves a vertical scar extending from the base of the tree to the height when the log first became lodged. Field observations suggest that a log may occasionally remain by the tree and abrade the tree during subsequent floods. Impact scars are generally more common along high-gradient streams and abrasion scars are more likely found in trees growing along low-gradient streams. However, abrasion scars sometimes develop when trees are impacted by unusually high velocities even during large-magnitude floods (Figure 5.4).

Numerous studies have used scars on riparian trees to estimate the crest stage of floods. Implicit in these studies is the assumption that scarring is most probable at or near flood crests. To test this assumption, the maximum heights of abrasion scars relative to the peak crest of a flood in 1990 was determined from 48 trees growing along the Skeena River in British Columbia, Canada. The mean height of scars was only 0.2 m lower than the actual flood crest. Along Buffalo Creek, a high-gradient stream in Colorado, the heights of impact scars from 102 trees along an 8 km reach were measured following a catastrophic flood in 1996. Of 67 trees that grew along the lowest parts of the floodplain, the average height of scars was 23 cm above well-defined crest elevations measured on the stream bank. Of 35 trees growing at higher flood-plain elevations, the average height of scars exceeded the flood crest by 17 cm. Thus, despite hydrologic differences between the two streams, heights of scars provided good estimates of actual flood-crest elevations. These studies strengthen the earlier findings of

Figure 5.4   An abrasion scar on the upstream-facing trunk of a Virginia pine (*Pinus virginiana* Mill.) from a great flood on the lower Potomac River in June 1972. The scar formed near the bottom of the trunk to a height of approximately 1.7 m. The top of the scar is about 10 m above the elevation of the mean river flow and is believed to mark the crest of the flood. The debris that abraded the bark has since rotted away. New growth is evident along the margins of the scar but it is still clearly visible more than 45 years after the flood. The bag is a reference for size. Photo by T. M. Yanosky.

Harrison and Reed that the maximum heights of scars on trees that were growing along the Turtle River in North Dakota were reliable proxies for flood crests determined from gauging records.

ABRUPT CHANGES IN RADIAL GROWTH

As mentioned previously, the expected trend of tree-ring widths is a gradual decline with increasing tree age. Chapter 4 discussed abrupt changes in ring width from one year to the next, but in these instances, typical radial growth subsequently resumes. Even most droughts do not result in reduced radial growth for more than a few consecutive years, particularly in regions such as the humid eastern U.S. However, sequences of unusually narrow rings, sometimes persisting for decades, often form following extreme damage to the roots or aerial parts of a tree. If the tree survives and eventually resumes an expected growth trajectory, the narrow sequence of rings provides not only the means to determine the onset of damage but perhaps may also be used as a rough estimate of the magnitude of damage. For example, the USGS located an ash tree that began to grow in 1870 immediately downstream from a rocky outcrop along the Potomac River near Washington, D.C. On the basis of comparable ring widths of 20-year-old trees that were presently growing nearby, the tree was probably about 5 m tall and its crown was likely well above the height of the rocky outcrop that sheltered the tree from damaging flood flows that occur, on average, every two years or less. The average width of rings from 1870 to 1889 was about 2.0 mm. A flood in early June 1889, of a magnitude comparable to the flood of record in 1936, apparently sheared off the crown, and radial growth from 1890 to 1920 averaged only about 0.4 mm before gradually recovering. We have observed reduced growth in numerous flood-damaged trees, but seldom to this extent. Further study is needed to determine the extent to which the degree and duration of reduced growth are correlated with the degree of damage, and whether either can be used as a proxy for the magnitude of the causative event.

Some abrupt changes in the rate of radial growth persist for most or all of the remainder of a tree's life. A familiar example is the enhanced growth of trees following the logging of neighboring vegetation, a *release* presumably due to reduced competition. Persistent environmental stress can likewise cause radial growth to decline for long periods and even result in tree death. A tree growing over a shallow aquifer, for example, may be negatively affected when contaminants transported in groundwater contact the root system. The USGS investigated a site in southern Massachusetts where an aquifer was contaminated by the surreptitious release of chlorinated hydrocarbons directly onto the surface of the ground. Water samples from wells confirmed the spatial extent of groundwater contamination, but the onset of contamination could not be determined. However, several black oaks (*Quercus velutina* Lam.) that grew along the contaminated flow path showed a synchronous and persistent growth decline compared to nearby trees growing outside the zone of contamination (Figure 5.5). It was speculated that dechlorination of hydrocarbons released large concentrations of chloride that caused the growth decline. Although trees showed no visible signs of ill health, element analysis of rings showed large concentrations of chloride in rings formed about 1970 and thereafter, which also coincided with declines in radial growth. Because chloride moves at the same approximate

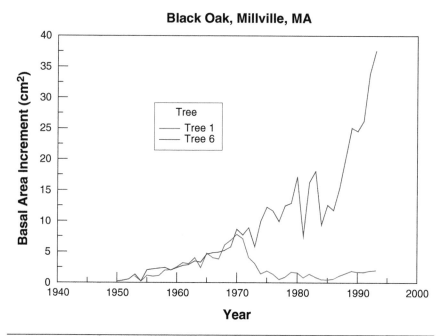

Figure 5.5   Ring-area growth, expressed as basal area increment, of two black oaks (*Quercus velutina* Lam.) growing in Millville, Massachusetts. Tree 1 is typical of trees growing along the flow path of shallow groundwater heavily contaminated by chlorinated hydrocarbons. Radial growth declined precipitously in the early 1970s when contaminated flow was believed to have first intersected the roots. Tree 2 grows just outside the contaminated zone. Note that trees are approximately the same age and that growth trajectories prior to the early 1970s are comparable.

velocity of groundwater, and because the release site could be determined with certainty, it was concluded from tree-ring evidence that the illegal dumping of hydrocarbons probably occurred in the late 1960s or very early 1970s. This raises the interesting possibility that the radial growth of trees might be used in lieu of groundwater monitoring as an inexpensive way to delineate the spatial extent and historical movement of contaminated groundwater.

Episodes of reduced radial growth alternating with periods of *normal* growth similarly can be used to study environmental factors that vary over time. Salt mining in central New York impacted the growth of white pine (*Pinus strobus* L.) trees growing in a small wetland over a period of 100 years. Salt was lifted from caverns 300–425 m beneath the ground by injecting water through steel casings into the caverns and pumping the brine to the surface. Some of the injected water was lost to the surrounding shales and moved along bedding fractures and into the wetland about 7 km away. The oldest pine in the wetland grew vigorously from the early 1860s until about 1889, at which time growth declined precipitously until about 1960. This precise period coincided with the use of injected water to lift the brine, and probably resulted in oversaturated conditions within the wetland that caused the 70-year period of reduced growth. In the late 1950s, however, water injections were discontinued, although brine continued to be lifted from the caverns. The wetland became drier as a result and the oldest tree resumed growth comparable to that prior to 1889. When mining ceased altogether in the late 1980s, however, the buildup of brine in the caverns resulted in increased flow into the wetlands. Unlike during earlier periods, the flow contained concentrations of salt sufficient to curtail growth and even kill some trees. Thus, trees in the wetland preserved evidence of long-term, episodic changes in hydrology that otherwise would be unobtainable.

If these hydrologic interpretations from botanical evidence are correct, it was hypothesized that all trees germinating during the 1889–1960 interval would grow slowly and resume a greater growth trajectory after 1960; similarly, trees that began growth between 1960 and the late 1980s would be expected to show vigorous radial growth. Indeed, these were the findings from 17 white pines sampled from the wetland. Additionally, selected trees were analyzed to determine the ring concentrations of sulfur and chloride; as expected, concentrations were greatest during the period that rings were most likely irrigated with brine (during the 1980s).

CHANGES IN RING ANATOMY

The types of cells and their arrangement and dimensions impart the characteristics by which the secondary xylem (*wood*) can be identified at least to genus. Although wood architecture is controlled primarily by genetics, environmental variables sometimes result in subtle anatomical changes that can be identified and even measured. Departures from expected anatomical form have primarily been used to describe the properties of wood for industrial use, including features that require chemical analysis such as variations in lignin and cellulose content. The alterations in ring anatomy being discussed here are confined to features that can be observed by using standard microscopy, although advances in image analysis technology may very well provide the means to objectify features that otherwise might be overlooked.

The wood of conifers is composed primarily of tracheids and, as a generalization, cells forming the earlywood have larger lumens than those in latewood. The transition from earlywood to latewood may be abrupt or gradual depending on the conifer species. Densitometric analysis using X-rays has been used to correlate ring density with environmental variables, but this requires expensive equipment beyond a simple microscope. The anatomical feature in conifers that is most useful to document environmental stress is the resin duct, a feature typically found in spruces (*Picea* spp.), pines, and larches (*Larix* spp.). These canals are lined with parenchyma cells that produce a resin important to the healing of wounds. Fires, frosts, insect infestations, and mechanical damage typically induce an increase in the formation of traumatic resin ducts. Some species that otherwise do not produce resin ducts will do so if injured. For example, the eastern hemlock (*Tsuga canadensis* (L.) Carr.) has been shown to produce traumatic resin ducts in response to an insect infestation (the wooly adelgid, *Adelges tsugae*).

The more complex ring architecture of hardwoods can be subdivided into ring porosity and diffuse porosity. Ring-porous species—which includes oaks, hickories, and ashes—produce large vessels in the innermost part of a ring and smaller ones thereafter. As a result, rings of these species generally are easy to delineate and false rings are unlikely. Diffuse-porous species—such as maples, birches, and gums—form rings in which the size of vessels is fairly uniform from the inner to the outer parts of the ring. It is often difficult to distinguish individual rings of these species, and great care must be taken to identify false rings prior to ring-width measurements.

The USGS surveyed damage to vegetation along several streams in the lower Potomac River basin following the catastrophic flood of late June 1972. Many trees formed a second crop of leaves shortly after most of the original foliage was stripped by flood waters. Numerous ash trees formed a row of new vessels in the 1972 latewood that resembled those produced during the first flush of radial growth in early spring (Figure 5.6). In other words, the production of spring-like vessels occurred within the latewood as the new crop of leaves expanded and matured. Generally, anomalous vessels were larger and more numerous in the upper parts than the lower parts of the trunk. These *flood rings* developed around the entire circumference of the 1972 ring, particularly near the apex of the trunk, suggesting that their formation is controlled by growth regulators that are transported basipetally. A question for further study is whether vessel morphology that is studied from the lower trunks of numerous trees can be positively correlated with the degree of flood damage and, thus, to flood magnitude.

To test the hypothesis that anomalous vessels developed only when a new leaf crop was produced, as opposed to a response to mechanical or physiological damage to the cambium, we hand-stripped the leaves from small ash trees and found that subsequent radial growth developed the same kind of anomalous vessels observed in flood-damaged trees. Increment corings and cross sections collected from additional ash trees located anomalous-vessel morphology in numerous years in which floods occurred during the growing season. An

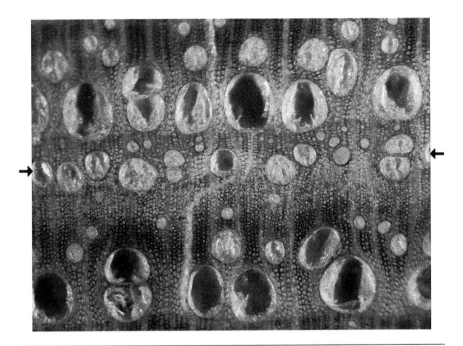

Figure 5.6 Transverse section of the 1972 and part of the 1973 rings of an ash (*Fraxinus* spp.) damaged by a great flood on the Potomac River in late June 1972. Direction of growth is from bottom to top. Leaves stripped by floodwaters were replaced by a new crop of leaves soon after the flood. Refoliation resulted in the production of a rank of anomalous vessels, or flood ring (arrows), in the outer part of the 1972 ring. See text for details. Magnification x70. Photo by T.M. Yanosky.

interesting finding was that the position of flood rings relative to ring width generally correlated with the timing of floods. A flood in late April or early May typically resulted in flood rings positioned in the early parts of rings; flooding late in the growing season was usually associated with flood rings in the outermost latewood. Therefore, we concluded the position of flood rings could be used to estimate the time of flood occurrence to within about two weeks. Altogether, anatomical evidence was found for 12 floods that preceded gauging records for the lower Potomac (1930), as well as three that occurred before records were kept at Point of Rocks (1895).

The earlywood of some ash trees contained vessels that were somewhat smaller than expected and that often grew in jumbled, less-ordered groups. Examination of flow records showed that these trees were flooded during periods of dormancy. It seems reasonable that moderate damage to shoots and buds might result in less hormonal activity upon the initiation of growth in the spring, which in turn might cause vessel morphology to be compromised. However, this explanation remains hypothetical and awaits further study.

Hypothetically, any ring-porous species would be expected to form anomalous vessels when defoliated and subsequently produce a new leaf crop. Oaks of several species grow along the Potomac River on rocky terraces but, unlike ash, are not prevalent at lower elevations. Some oaks attain ages exceeding 200 years and have probably been flooded during unusually large events, but to our knowledge they have not been examined for flood-induced ring growth. Swamp white oak (*Quercus bicolor* Willd.) is rare but occasionally is found on the frequently flooded bedrock just upstream from Washington, D.C., but again, we are unaware of investigations into their ring morphology.

In addition to the formation of anomalous vessels following flood damage, the morphology of fibers in the rings of ash trees sometimes changes when roots are flooded during the growing season. Fibers typically develop larger lumens and thinner walls, often resembling fibers formed at the very beginning of growth in the spring. Anomalous fibers often appear as bands or *white rings* that correspond to known episodes of root flooding that otherwise do not seem to damage the tree. Rather, flooding for short periods (one to three days) seems to accelerate the rate of growth. We have observed several bands of anomalous fibers within a single ring, each seemingly having developed as the result of known episodes of root flooding. As expected, rings with multiple bands of spring-like fibers are typically quite wide, strengthening the contention that short-term root flooding accelerates the rate of radial growth. Again, there seems to be ample opportunity for additional basic and applied research.

The finding that flooding can alter the ring anatomy of ash trees begs the question of whether diffuse-porous species—commonly found in wetlands and floodplains—likewise develop growth alterations following flood damage or episodic inundation of roots. The more uniform structure of diffuse-porous rings presents both a problem and an opportunity; whereas flood rings in ring-porous species do not have the appearance of actual false rings, diffuse-porous species are notorious for producing false rings that are difficult to differentiate from true annual rings. However, the very presence of false rings is evidence that diffuse-porous species produce altered ring morphology in response to environmental variables. Thus, the variable that should be studied is flooding. How do diffuse-porous species respond to damaging floods? Or to episodes of periodic root flooding? Is one species a better *recorder* than others, and if so, under what range of conditions? Some relevant research has been conducted but up to now the proverbial surface has hardly been scratched.

A final consideration is the use of wood anatomy to document periods of stem burial (deposition) or the excavation of roots by surficial erosion. The literature contains assertions that buried stemwood subsequently develops anatomical features that are characteristic of roots and that exposed roots become more stem-like. However, considerable debate exists on this matter, and it may not be farfetched to conclude that resolution of the question resides in the mind of the observer. On the other hand, given the large numbers of tree species and their variable abilities to respond to a vast continuum of environmental stresses, it seems reasonable to expect that there is no consistent growth response and that cases should be resolved from species to species.

I (TMY) was once sent cross sections from the roots and trunk of a red maple (*Acer rubrum* L.) that grew in central New York where subterranean mining caused localized surface subsidence. The sender wondered if I could determine when the roots became exposed; he believed that subsidence had occurred quickly and had a good idea when it occurred as well. I was unable to note any anatomical differences between the root wood and stem wood, but I noticed that each ring from the trunk contained small deformations caused by burrowing insect larvae (*pith flecks*). The root wood also contained pith flecks, but only in the outer rings. None were observed in rings that formed prior to 1946, so I tentatively concluded that 1946 might be the year when subsidence occurred. I received an e-mail from the sender saying that, on the basis of a number of different pieces of evidence, he had previously concluded that 1946 indeed was the year of subsidence. This could have been a coincidence, of course, but it would be interesting to see if pith flecks have utility in ring-anatomical studies since they are typically common in the wood of diffuse-porous trees.

SOME CONCLUDING REMARKS

This chapter began a basic discussion of the responses of trees to environmental factors as determined from their annual rings. These factors may be events of short duration that recur at irregular intervals and vary in magnitude or they could be processes that occur continuously throughout the life of a tree. Considered together, radial-growth responses are adaptations that promote survival, and thus are necessary to better understand the total ecological environment of trees. Woody vegetation growing on an upland, in a wetland, or along a stream may differ greatly in terms of species composition that results from adaptations to local environmental pressures. Because some trees attain great age and their rings usually can be precisely dated, ecological inferences can sometimes be made over long periods, in turn permitting determinations of environmental stability.

Although presented separately, researchers typically use all applicable techniques when approaching problems in the field. For example, the studies mentioned earlier of mudflows on Mt. Shasta and the Cascadia earthquake each used data from trunk scars, reaction wood, ring-width changes, and ring-anatomical responses (in both cases, the presence of traumatic resin ducts). Robert Sigafoos used scar data, the age of sprouts, and reaction wood in his pioneering studies of floods and flood-plain vegetation. (As an interesting aside, he often referred to his studies as *simple-minded botany* because it was based on seemingly elementary principles of tree growth—to him, a Harvard-trained PhD, at least.)

Sigafoos' humor notwithstanding, the researcher quickly learns that in actuality this science is far from easy. To anthropomorphize a bit, trees have a nasty habit of not being particularly cooperative. Trees at a study site may not be present or be of insufficient age; or they simply may not respond in a manner that preserves the evidence that is sought. Every dendrochronologist has located a tree that he is certain is the key to his field problem, only to find that the trunk is hollow. Similarly, trees may contain so many false and missing rings that it is impossible to assign precise dates. Thus, all studies are constrained by the availability of pertinent study material and, thereafter, by the cleverness of the investigator.

We are often asked about the minimum number of samples required to reach reasonable conclusions. Unlike statistically based reconstructions that require dozens of trees, often from more than one site, the methods discussed in this chapter often rely on evidence from only a few trees. Large, infrequent events such as the 1972 flood on the Potomac River impacted vast numbers of trees, and thus there was no dearth of trees to sample. With time, however, evidence becomes less common because trees die, and the occurrence of an even larger flood sometimes destroys evidence of earlier floods. It is the prerogative of the investigator to decide whether conclusions are warranted from the evidence at hand. Many researchers are willing to draw conclusions concerning historical events from only one or two very old trees, but only if similar evidence for more recent events can be found.

Unusually large or catastrophic events often destroy evidence of earlier events, sometimes over large areal extents. For example, Don Miller of the USGS studied the generation of giant waves in Lituya Bay, Alaska. He found that rock and ice falls from cliffs at the head of the Bay created giant waves that splashed against the surrounding topography. He found botanical evidence for waves generated in 1853, 1874, 1899, and 1936. However, an unusually large rock fall in 1958 produced a wave that reached approximately 524 m above sea level and destroyed all vegetation in its path, including all evidence of earlier waves. In similar fashion, we have rarely found botanical evidence for floods on the Potomac River that occurred in 1936 (the flood of record) and 1937, even though evidence must have been extremely frequent at the time. The great flood of October 1942 was slightly less than that in 1936, but apparently this flood obliterated evidence of the two earlier large floods. Interestingly, the June 1972 flood, although smaller than that in 1942, nevertheless destroyed several trees that preserved evidence of the 1942 event. Thus, a researcher seeking botanical evidence of extremely large, historical floods (or any event expected to impact vegetation) should survey trees growing near the crest elevation of the flood. I (TMY) have often wondered if evidence persists from the great Johnstown flood of late May 1889, as it has thankfully never been exceeded.

The persistence of evidence for environmental factors affecting tree growth depends on the nature of the evidence and the life span of trees. Anatomical features, including reaction wood, traumatic resin ducts in conifers, and abrupt changes in ring widths all are preserved until the tree dies. Even though scars eventually heal over, Alex McCord of the University of Arizona took increment corings at successive trunk heights and was able to locate subsumed scars permitting a reconstruction of flood histories back to 1471. In the humid regions of the eastern United States, the persistence of botanical evidence is doubtless much shorter. Our colleague Cliff Hupp at the USGS recovered evidence of a flood on Passage Creek in northwestern Virginia that occurred in 1720, but the ages for most historical floods are much younger.

This brief survey of botanical methods used to understand the ecological environment of trees can be expanded in much greater detail by consulting the wealth of peer-reviewed literature published in the last several decades. In addition to the kinds of events discussed here, tree-ring studies also have been applied to earthquakes, wildfires, volcanic activity, and even avalanches.

SELECTED REFERENCES

Gottesfeld, A. S. (1996). "British Columbia flood scars: Maximum flood-stage indicators." *Geomorphology* 14: pp. 319–325.

Harrison, S. S. and J. R. Reid, (1967). "A flood frequency graph based on tree scar data." North Dakota Academy of Science, Proceedings, pp. 23–33.

Hupp, C. R. and E. E. Morris, (1990). "A dendrogeomorphic approach to measurement of sedimentation in a forested wetland, Black Swamp, Arkansas." *Wetlands* 10: pp. 107–124.

Jacoby, G. C., D. E. Bunker, and B. E. Benson, (1997). Tree-ring evidence for an A.D. 1700 Cascadia earthquake in Washington and northern Oregon." *Geology* 25: pp. 999–1002.

McCord, V. A. (1996). Fluvial process dendrogeomorphology: Reconstructions of flood events from the southwestern United States using flood-scarred trees" in *Tree Rings, Environment and Humanity,* edited by Dean, J. S., D. M. Meko, and T. W. Swetnam, *Radiocarbon,* University of Arizona, Tucson, AZ. pp. 689–699.

Miller, D. J. (1960). "Giant waves in Lituya Bay, Alaska." U.S. Geological Survey Professional Paper 354-C. pp. 51–86.

Sigafoos, R. S. (1964). "Botanical evidence of floods and flood-plain deposition." U.S. Geological Survey Professional Paper 485-A. p. 35.

Stoffel, M., et al., editors. (2010). *Tree Rings and Natural Hazards: A State-of-the-Art.* Springer. p. 505.

Wieczorek, G. F., et al. (2006). "Hurricane-induced landslide activity on an alluvial fan along Meadow Run, Shenandoah Valley, Virginia (eastern USA)." *Landslides* 3: pp. 95–106.

Yanosky, T. M. (1983). "Evidence of floods on the Potomac River from anatomical abnormalities in the wood of flood-plain trees." U.S. Geological Survey Professional Paper 1296. p. 42.

Yanosky, T. M. and R. D. Jarrett, (2002). "Dendrochronologic evidence for the frequency and magnitude of paleofloods." *Ancient Floods, Modern Hazards: Principles and Applications of Paleoflood Hydrology, Water Science and Application* 5: pp. 77–89.

Yanosky, T. M. and W. M. Kappel, (1997). "Effects of solution mining of salt on wetland hydrology as inferred from tree rings." *Water Resources Research* 33: pp. 457–470.

Yanosky, T. M., B. P. Hansen, and M. R. Schening. (2001). "Use of tree rings to investigate the onset of contamination of a shallow aquifer by chlorinated hydrocarbons." *Journal of Contaminant Hydrology* 50: pp. 159–173.

6

BASAL AREA INCREMENT AND THE NON-CLIMATIC COMPONENT

Much of our discussion of tree rings has been with regard to ring widths measured on a transverse surface. In Chapter 3, we described measurement of tree rings, and then standardization of the rings into tree-ring indices. In Chapter 1, we described tree rings as expressions of 3-dimensional annual increments, and pointed out that ring widths only describe one of the three dimensions. In this chapter, we will turn our attention to describing rings in two dimensions, as they might appear on a transverse section. The descriptive term that we will use is basal area increment (BAI). We will start with some calculations of BAI and then describe some applications and interpretations of results.

Basal area (BA) is a cross-sectional area of a tree stem at basal height (breast height). Foresters and ecologists have long utilized BA as a measure of tree size. At one time, in trying to explain to a forester something that we were doing with ring area, we commented that ring area could be thought of as just an annual increment of basal area. From that we coined the term BAI—or so we thought. It turns out that BAI is a term that is not original with us at all. It has been used in forestry, though in a somewhat different application, for quite a few years. For example, stand BA is often determined from diameter at breast height (dbh) measurements. If the stand were to be re-sampled several years later, the difference in stand BA between measurement periods can be referred to as BAI. On the other hand, we have not run across anything suggesting BAI was previously used to refer to annual increments calculated from measurements of tree rings. The forestry use of the term BAI does not usually refer to annual increments. Perhaps, in tree-ring applications referring to annual increments, it would be more accurate to use a distinguishing term such as annual increment of basal area (AIBA or ABAI). On the other hand, BAI, as applied to tree rings has now appeared in the literature; thus, rather than try to introduce another new term, we will continue to use the term BAI to refer to annual increments as calculated from tree-ring measurements.

BAI is simply jargon for tree-ring area. Strictly speaking, BAI should refer only to ring area at basal height. A more generalized term that we could use to refer to an annual increment of ring area at any given height is ring-area increment, RAI. Because the greatest majority of tree-ring work is done with increment core samples taken at basal height, BAI is expected to remain the more commonly used term. In the following discussions we will use BAI throughout, though in some cases RAI might be a more correct term.

BASIC BAI CALCULATIONS

The clear, branch-free portion of a trunk, or bole, of a canopy-sized deciduous tree contained branches when the tree was smaller. Once the tree reached canopy stature, however, the branch-free portion was pretty well set. For hardwood species, the branch-free portion is often referred to by foresters as the marketable bole. For the rest of the life of the tree, the diameter of the trunk and the height of the tree may increase considerably, but the length of the branch-free portion of the tree trunk will normally change very little. Often, the lower branches of the crown develop into major branches and are not expected to die until the tree reaches what foresters refer to as maturity.

The shape of the main trunk of a tree may be thought of as approaching that of a paraboloidal segment. Ring area of an individual ring of the paraboloidal model described in Chapter 1 was described as remaining constant with height. Volume of an individual ring of a tree trunk segment can thus be determined by simply multiplying ring area (BAI) by trunk segment length.

BAI of the nth ring can be calculated using an equation that is a restatement of the ring area equation (Eq. 1.7):

$$BAI_n = \pi\left(r_n^2 - r_{n-1}^2\right)$$ (Eq. 6.1)

where

r_n = radius of the circle describing the outside boundary of the nth ring, and
r_{n-1} = radius of the circle describing the outside boundary of ring $n-1$.

If we factor the BAI equation (Eq. 6.1), we find that:

$$BAI_n = \pi(r_n + r_{n-1})\,(r_n - r_{n-1})$$

but, of course,

$(r_n + r_{n-1})$ = mean diameter at the nth year, or dbh, and
$(r_n - r_{n-1})$ = ring width at the nth year.

Thus, we can think of BAI of the nth year as being π times mean dbh times ring width at the nth year. In other words, BAI can be thought of as ring width $(r_n - r_{n-1})$ adjusted for tree size (dbh).

The area of a circle is π times the square of the radius. If we assume that tree-ring boundaries approximate circles, then the area of a tree ring is simply the difference between consecutive circles defined by ring boundaries. To calculate an estimate of ring area for a series of tree rings, we need a set of radii composed of radius length to the outside of each ring. Radius length, but not ring width, is included in Equation 6.1. In practice, ring width may be used to calculate radius length. One could start with the center ring and add ring widths to obtain such a set of radii. The radius length to the outside of the first ring is simply the width of the first ring, and the radius length to the outside of the second ring is the sum of the first two ring widths, and so forth.

On the other hand, it may be more practical to calculate radii from total radius length. Thus, the set of radii is obtained by subtracting successive ring widths from total radius length. In practice, total radius is determined by any of the following approaches.

If the wood sample (usually an increment core) includes the pith, and the wood sample has been measured to the pith, then simply use the sum of all ring widths for total radius length.

If the sample, taken from the outside, does not quite reach the pith, then the radius of the missing inner rings can be estimated with an mm ruler from the curvature of the arc of the innermost included ring. Rather than add this estimate to the total of the measured ring widths, it may be just as easy to directly measure total radius with the mm ruler.

If the wood sample includes only a small part of the total radius, or there is not enough curvature to the inner included ring to allow a reasonable estimate of the missing radial segment, then there may be little choice but to estimate radius length from a field measurement of dbh. Remember to allow for bark thickness; thus, radius length would be $\frac{1}{2}dbh_{ib}$.

Because the botanical center (pith) may be offset from the geometric center, check to make sure that the radius estimated from dbh is equal to or greater than the sum of the measured widths; otherwise, some of the center rings could end up being calculated as negative BAI values—not the end of the world, but inconvenient.

CLIMATIC AND NON-CLIMATIC COMPONENTS

A ring-width series was described in Chapter 3 as being composed of two components: a climatic component (year-to-year variation) superimposed on a non-climatic component (time trend). The non-climatic component was estimated by fitting a smooth curve to a raw ring-width series. Ring-width indices were described as an estimate of the standardized climatic component. Indices were obtained by dividing raw ring-width values, w_r, by corresponding values, w_s, from the smooth curve that was used to represent the non-climatic component (Eq. 3.4). In effect, calculating indices amounts to identifying the non-climatic component and then getting rid of it by dividing it out.

Ring-width data can be transformed to area data while still maintaining the component concept (Figures 6.1–6.4). Raw basal area increment, BAI_r, may be calculated from radius lengths with Equation 6.1. Smoothed BAI_s, can be calculated with Equation 6.1 by using smoothed ring widths, w_s. In theory, BAI_s could also be obtained by fitting a curve to the BAI_r values. In practice, BAI_s smoothed from BAI_r is identical to BAI_s transformed from w_s only when the curve-fitting process is applied with equal accuracy throughout the time series. Usually, the curve fitting is most accurate for the early years (wider rings) in the ring-width series, and for the later years (greater BAI values) in the BAI series. We found that there was a chance of encountering negative calculated values when dealing with outside rings of a large tree showing a recent sharp growth decline. If the decline was sharp enough, then smoothing BAI_r could result in negative BAI values for the last one or more rings. For this reason, we routinely smoothed the ring-width series rather than the BAI series.

Indices (Figure 6.4) can be calculated either by dividing raw ring-width values, w_r, by corresponding values, w_s, from the ring-width curve, or by dividing BAI_r by BAI_s. Theoretically, area indices calculated from area data (BAI) should be identical to ring-width indices (I) calculated from ring-width data (w). Again,

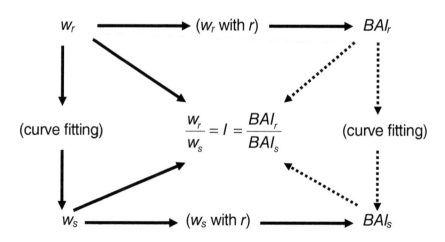

Figure 6.1 Flowchart indicating transformation pathways from ring-width (w) to ring area (BAI) and indices (I). Solid arrows indicate preferred pathways.

(a)

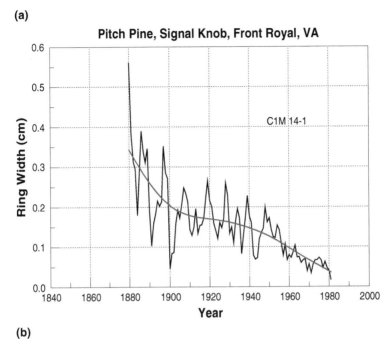

(b)

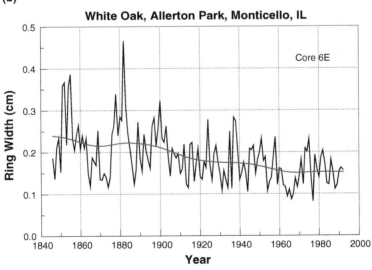

Figure 6.2   Raw ring width (w_r in Figure 6.1) and smoothed ring width (w_s in Figure 6.1). Shown are examples of (a) a softwood, pitch pine (*Pinus rigida* Mill.) collected at Signal Knob, Front Royal, VA, by John Whiton of the United States Geological Survey (USGS) Laboratory of Tree-Ring Research (LTRR), and (b) a hardwood, white oak (*Quercus alba* L.) collected at Allerton Park, Monticello, IL, by a dendrochronology class at the University of Illinois.

(a)

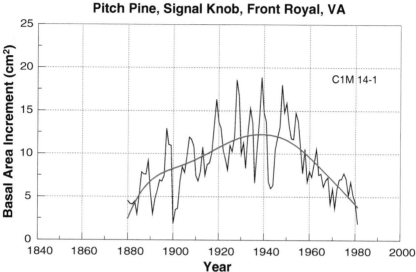

(b)

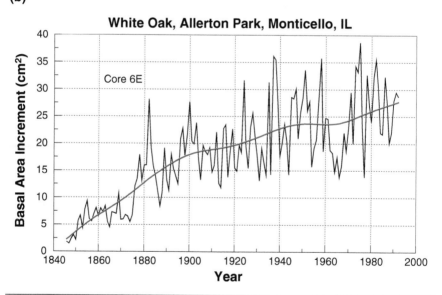

Figure 6.3   Raw BAI$_r$ of Figure 6.1 and smoothed BAI$_s$ of Figure 6.1. Data are of (a) Signal Knob pitch pine, C1M 14-1, and (b) Allerton Park white oak, 6E. Note marked growth decline in pine thought to be induced by point source pollution (viscose rayon plant in nearby Front Royal that began operation in 1939). Oak reached canopy size circa 1900.

(a)

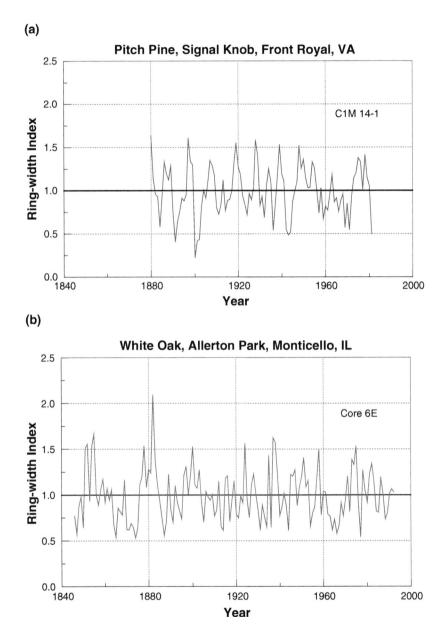

(b)

Figure 6.4   Ring-width indices, *I*, of (a) Signal Knob pitch pine, C1M 14-1, and (b) Allerton Park white oak, 6E. The preferred method of calculating indices is to divide raw ring width, w_r, by smoothed ring width, w_s (Figure 6.2). Indices may also be calculated by dividing BAI_r by BAI_s (Figure 6.3), although the degree to which the results of the two methods agree is dependent on the accuracy of the smoothing function.

though, because of curve-fitting accuracy, the two procedures often produce results that are not precisely identical.

The smoothed ring-width series (Figure 6.2) has been described as an estimate of the non-climatic component of ring width. The smoothed area curve (Figure 6.3) may be regarded as an estimate of the non-climatic component of the ring-area (BAI) series. An important point to re-emphasize here is that though the non-climatic component of ring width is typically curvilinear, particularly in deciduous trees that have been in the canopy for 200 or fewer years, transformation to BAI results in a quasi-linear non-climatic component for years that the trees have been in the canopy.

THE HYPOTHETICAL STEM REVISITED

Figure 1.12 is shown again in this chapter as Figure 6.5 because of its importance to the discussion. Because height growth in Figure 6.5 is modeled as a constant, ring-area (BAI), it is the same at any ring and any node. Table 6.1 includes ring-width data for all rings at all nodes shown in Figure 6.5, exclusive of the apex (also a node), and are based on arbitrarily assigning the center ring a width of w. Reading the table horizontally (by row) gives the chronological ring-width sequence that could be obtained from an increment core or a transverse section.

Reading the table vertically (by column) also gives the ring-width variation by year, but all are of the same ring number from the center. Ring widths among cells of each vertical sequence are equal. Reading the table diagonally gives the width variation within a given ring; hence, each sequence is for a single year. It was pointed out in Chapter 1 that the width of a given ring decreases basipetally (diagonally in Table 6.1) much as ring widths decrease with age at a given height (horizontally, Table 6.1 and Figure 6.3). Actually, according to the model (Figure 6.5), proceeding basipetally from the apex of any given ring, ring width increases to the first node below the apex, then decreases. Ring widths of Table 6.1 are represented graphically in Figure 6.6.

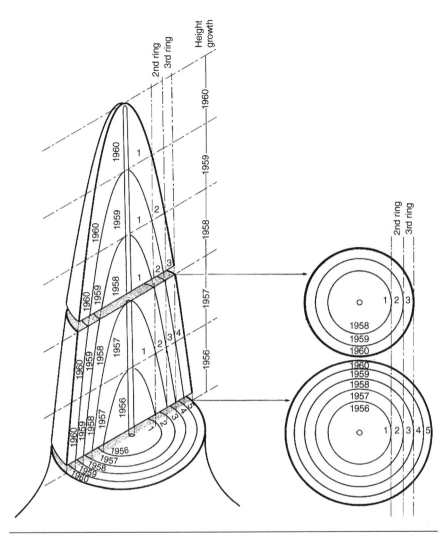

Figure 6.5 Hypothetical stem composed of five paraboloidal growth layers (tree rings). Geometrically, the outer surface of each successive growth layer simply describes a longer segment of the same paraboloid. Because height growth (increase in segment length) has been held at a constant in this model, BAI is the same for all rings at all nodes. Transverse sections shown are for the first and third nodes acropetally from the base.

Table 6.1   Ring widths at each node of the hypothetical stem (Figure 6.5) based on a perfect set of paraboloidal segments. [To produce the table, the inside ring at each node was assigned a value of w. Dates of ring formation, as shown in Figure 6.5, are in parentheses. Nodes are numbered in chronological order, acropetally from the base of the stem.]

Node No.	Radius Sq. (r^2)	Total Radius	Ring Widths (at Nodes)				
5	w^2	w	w (1960)				
4	$2w^2$	$\sqrt{2}\cdot w$	w (1959)	$w\left(\sqrt{2}-1\right)$ (1960)			
3	$3w^2$	$\sqrt{3}\cdot w$	w (1958)	$w\left(\sqrt{2}-1\right)$ (1959)	$w\left(\sqrt{3}-\sqrt{2}\right)$ (1960)		
2	$4w^2$	$2w$	w (1957)	$w\left(\sqrt{2}-1\right)$ (1958)	$w\left(\sqrt{3}-\sqrt{2}\right)$ (1959)	$w\left(2-\sqrt{3}\right)$ (1960)	
1	$5w^2$	$\sqrt{5}\cdot w$	w (1956)	$w\left(\sqrt{2}-1\right)$ (1957)	$w\left(\sqrt{3}-\sqrt{2}\right)$ (1958)	$w\left(2-\sqrt{3}\right)$ (1959)	$w\left(\sqrt{5}-2\right)$ (1960)

Perhaps of greater interest at this point are sequences of area (BAI) and volume. BAI values, calculated from the ring widths at each node (Table 6.1), are presented in Table 6.2. Again, ring width of the center ring at any node was arbitrarily assigned a value of w. The perfect set of paraboloids results in all BAI values being equal (Figure 6.7). Proceeding basipetally within any given ring, ring area (BAI) is zero at the apex, increases to πw^2 at the first node below the apex, and then remains constant. Also, except for the center ring at the apex, internode segment volume (Table 6.3) is also equal among rings ($\pi w^2 h$). Because BAI increases from zero at the apex in the terminal internode, the ring volume of the internode is only half what it is on lower internodes. In practice, increment cores often do not include the center ring but even if they do, there is no way of knowing where along the terminal internode a given sample core intersected the internode. The important point here is that BAI is directly proportional to volume (simply BAI times the constant, height). That the terminal leader does not follow the relationship of the rest of the stem is rarely of any practical consequence.

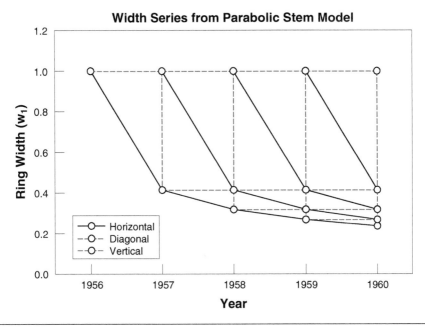

Figure 6.6 Ring-width series at each node of the hypothetical stem (Figure 6.5) based on data from Table 6.1 in which the center ring at each node is assigned a width of *w*. The row sequences (solid black lines) represent series that would be obtained from increment cores or transverse sections.

Table 6.2 BAI at each node of the hypothetical stem (Figure 6.5) based on ring widths from Table 6.1. [Years, as labeled in Figure 6.5, are in parentheses.]

Node No.	Radius sq. (r^2)	Stem Area	BAI (at Nodes)				
5	w^2	$\neq w^2$	$\neq w^2$ (1960)				
4	$2w^2$	$2 \neq w^2$	$\neq w^2$ (1959)	$\neq w^2$ (1960)			
3	$3w^2$	$3 \neq w^2$	$\neq w^2$ (1958)	$\neq w^2$ (1959)	$\neq w^2$ (1960)		
2	$4w^2$	$4 \neq w^2$	$\neq w^2$ (1957)	$\neq w^2$ (1958)	$\neq w^2$ (1959)	$\neq w^2$ (1960)	
1	$5w^2$	$5 \neq w^2$	$\neq w^2$ (1956)	$\neq w^2$ (1957)	$\neq w^2$ (1958)	$\neq w^2$ (1959)	$\neq w^2$ (1960)

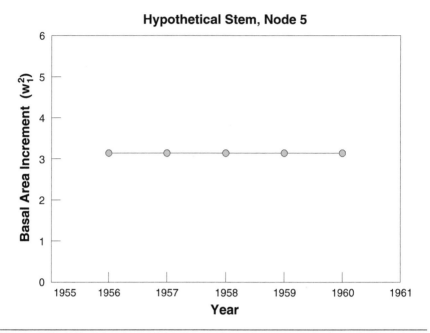

Figure 6.7   BAI of the rings of the first node at the base of the hypothetical stem (Figure 6.5); based on data from Table 6.2 in which width of the center ring is *w*, the slope of the BAI trend line is zero.

According to the growth model, ring width at any given height follows a die-away curve with wider rings in the center and more narrow rings toward the outside. Interestingly, the same pattern is shown at all heights, but the length of the curve (number of rings) decreases with increasing height (Figure 6.6).

GIVING BAI A POSITIVE SLOPE

The calculations from the hypothetical stem are based on all ring areas being equal (Figures 6.5 and 6.7). Thus, for the hypothetical model, the BAI trend has a slope of zero. In the real world, we find that the BAI trend is expected to have a positive slope and that the slope is normally related to site quality; that is, the better the site quality, the greater the slope. Faster growing white oaks

Table 6.3 Ring volume by internode of the hypothetical stem shown in Figure 6.5. [Based on ring widths from Table 6.1 and segment height (h). Internodes are numbered acropetally from the base. Internode 1 is from the base (node 1) to node 2. The apex is node 6. Years, as shown in Figure 6.5, are in parentheses.]

Internode No.	Radius Sq. (r^2)	Internode Volume	Ring Volume (by Internode)				
5	w^2	$\dfrac{\neq w^2 h}{2}$	$\dfrac{\neq w^2 h}{2}$ (1960)				
4	$2w^2$	$\dfrac{3(\pi w^2 h)}{2}$	$\dfrac{\neq w^2 h}{2}$ (1959)	$\dfrac{\neq w^2 h}{2}$ (1960)			
3	$3w_1^2$	$\dfrac{5(\pi w^2 h)}{2}$	$\dfrac{\neq w^2 h}{2}$ (1958)	$\dfrac{\neq w^2 h}{2}$ (1959)	$\dfrac{\neq w^2 h}{2}$ (1960)		
2	$4w_1^2$	$\dfrac{7(\pi w^2 h)}{2}$	$\dfrac{\neq w^2 h}{2}$ (1957)	$\dfrac{\neq w^2 h}{2}$ (1958)	$\dfrac{\neq w^2 h}{2}$ (1959)	$\dfrac{\neq w^2 h}{2}$ (1960)	
1	$5w_1^2$	$\dfrac{9(\pi w^2 h)}{2}$	$\dfrac{\neq w^2 h}{2}$ (1956)	$\dfrac{\neq w^2 h}{2}$ (1957)	$\dfrac{\neq w^2 h}{2}$ (1958)	$\dfrac{\neq w^2 h}{2}$ (1959)	$\dfrac{\neq w^2 h}{2}$ (1960)

from the rich, glacial Wisconsin till plain at Allerton Park in central Illinois, for example, show a steeper BAI slope (0.192) than do slower growing white oaks (slope of 0.069) from a hardscrabble mountain top at the Lilley Cornett tract in eastern Kentucky.

Calculations for area and volume, such as in Tables 6.2 and 6.3, get quite complex when the BAI trend is given a positive slope. We will simply graphically show results for BAI and ring width (Figures 6.8 and 6.9) for three slope values. In practical applications, BAI slope can be obtained from the regression equation of the time period of interest.

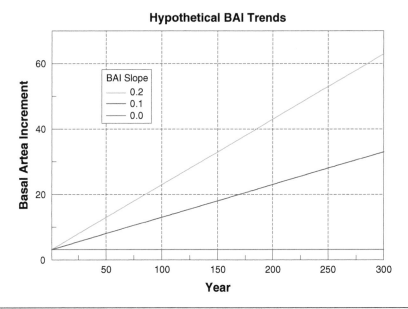

Figure 6.8　Simulations of BA series at three different yearly rates of increase. A series with no increase would have a slope of zero (0.0). Also shown are slopes of 0.1 and 0.2.

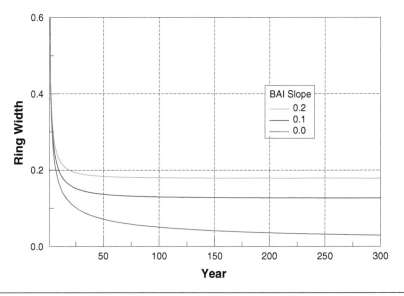

Figure 6.9　Simulations of ring-width series that would have produced the BAI trends shown in Figure 6.8. Note that as BAI slope increases, the length of time for ring width to reach a near constant value decreases.

THE EXPECTED LINEAR TREND OF BAI

It has been shown that if BAI increases only very slightly from year to year, geometric constraints result in a ring-width trend graphically resembling a die-away curve (Figure 6.9), particularly for trees less than 200 to 300 years old. We have also shown that there is a tendency for the BAI trend to be linear with a positive slope. Having examined hundreds of BAI trends from many deciduous species in many habitats leaves little doubt that in the eastern deciduous forest, the expected trend is near linear for the years that the tree is in the canopy.

In perhaps oversimplified terms, photosynthetic potential may be regarded as a function of crown diameter. If, in open-grown trees, growth in length of stems results in a fairly steady, linear increase in crown diameter, this might explain the linear BAI trend. A direct connection between crown diameter and BAI would go a long way in explaining why the BAI trend seems to reflect crown history. The story gets a little murky for trees in a closed canopy in which BAI continues to increase, though the rate at which crown diameter increases may be quite constrained as competition among trees increases. The relationships between crown size, photosynthetic potential, and BAI are certainly potential areas for additional work.

MERGING DATA

BAI may be thought of as ring width that has been adjusted for tree size; that is, for radius length. For this reason, ring-width data from individual samples should not be averaged together before calculating BAI; averaging, if done, should be done after BAI values have been calculated. That is to say, averaging should not be done until after adjustments for radius length have been made. This is particularly important if the inside dates of the samples are not all the same, as in samples from mixed-age stands.

An example using data from only two samples, samples 8e and 8w from tree 8 of the Allerton Park white oak collection is shown in Figure 6.10. In this example, calculating BAI from mean width data will inflate the BAI values.

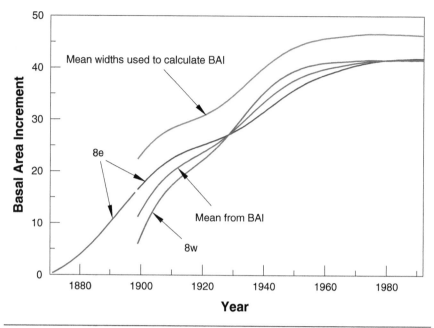

Figure 6.10   Contrasting BAI values calculated from mean-width data and BAI values calculated as the mean of BAI values for each sample.

INDIVIDUAL SAMPLE DATA VERSUS MEAN COLLECTION DATA

There may be considerable variation among trees even from a single species at a uniform site. For this reason, a single sample may not be very representative of a site. A better representation of a site may be indicated by averaging the samples from the site.

When we first put together a program for calculating BAI, the results (raw BAI, smoothed BAI, and Indices) were all given as mean collection values. Then we realized that sometimes it might be worthwhile to look at results from individual samples. We found that, as a general rule, the more variation there was in the site or in the site history, the less uniform the collection was, and the more we gained by looking at data from individual samples. Turning this around, sometimes we didn't know about the varied history of the site, but comparison of data from individual samples would provide some valuable clues.

As a result of this, we incorporated an option in our program that would allow us to look at ring width, raw BAI, smooth BAI, and indices from individual samples. On the other hand, in working with relatively uniform collections, we often didn't bother with the *individuals* option.

In working with data from individual samples, the results from a given sample were typically not as important as comparisons among samples. Sometimes the trends of just a few samples were controlling the mean. If the same pattern was showing up in most of the samples, it might be suggested that the trees were responding to something in common among trees. As one gains more knowledge of the trees, the site, and the site history, more things may become apparent in the story that the tree rings are capable of telling us.

INTERPRETATION OF BAI TRENDS

The BAI trend can be regarded as representing the history of the crown. The slope of the BAI trend can be regarded as a function of site quality; that is, the faster growing trees display a greater BAI slope. A slope of zero suggests that new growth in the crown is just equal to the amount of die back and self-pruning. If we think of growth rate as being the amount of growth per year, then the BAI slope may be regarded as the rate at which growth rate is increasing. BAI slope may also be thought of as an expression of the relationship between growth and environment (climate). An unchanging slope would usually indicate that the relationship between growth and environment is not changing enough to affect the rather steady rate at which the crown is increasing in size.

Taken together, crown history and site quality can provide useful information regarding growing conditions of a tree. We find that there is often an increase in slope of the BAI trend when a deciduous tree in an established closed-canopy forest attains canopy status. The crown is released from suppression and then increases, sometimes sharply. The opposite reaction, a decrease in BAI slope from an increase in crowding and competition, can happen when the crowns of open-grown trees become large enough that the crowns merge.

Conifers would be expected to behave more like open-grown deciduous trees. Thus, the increase in competition among tree crowns as the young stand closes in, would be expected to show a decrease in BAI slope (circa 1890 in pitch pine and circa 1900 in white oak, Figure 6.3).

Whether the BAI trend increases (release from suppression) or decreases (increase in suppression), the slopes and slope changes of understory trees can be quite a challenge. For this reason, many studies confine analysis to those years when a tree was in the canopy. On the other hand, the amount of information included in all of the BAI slope changes of the understory trees might provide considerable insight into understory stand dynamics.

Crown releases can, of course, also occur in established canopy trees. The first reaction would be an immediate increase in BAI slope. Then, because the relationship between growth and environment has not changed, after the crown has adjusted to the release, the slope would be expected to return to what it was before the release. In other words, a release would appear as an upward shift in the BAI trend, such that the amplitude of the shift reflects the degree to which the tree crown increased in size during the release.

SIMULATIONS OF LONG-TERM RING-WIDTH TREND

Determinations of total growth of a tree could be quite complex; however, what we are really concerned with is annual growth of the main stem in the form of tree rings. We are not particularly concerned with such things as phloem growth, root growth, or the production of new leaves. We don't have any way of knowing how these things varied in the past, and perhaps as a consequence, we do not have any good indication of how annual variation in these things may have affected variation in ring development. Even annual variations in leaf and fruit production, that are known to be interrelated with the size of the ring produced on any given year, have only received minimal attention in the literature. A comparison of BAI trends of apple trees in an orchard compared to the apple yield of that orchard, as simple as it sounds, has probably never been done.

The hypothetical stem (Figure 6.5) is based on ring area at any given height being constant; that is, the data (horizontal series of Table 6.2) would graph as a straight line with no slope (Figures 6.7 and 6.8). As already noted, real-world data for the years that the tree is in the canopy tend to be a straight line, but have a positive slope (Figure 6.11).

The Lilley Cornett collection (Figure 6.11) contains trees in excess of 300 years, yet in 1980 the BAI trend was continuing in a straight line. Growth quantity was still increasing each year. Obviously, that trend of increasing BAI cannot continue forever. On the one hand, if the BAI slope is zero (as illustrated in Figures 6.5 and 6.8), then we might expect that with increasing age, while maintaining a constant BAI, ring width would decrease as diameter got larger. If slope is positive, it would seem that ring width would not decrease as quickly as in the steady state scenario of no slope. Indeed, if the yearly increase in BAI were great enough (that is, if BAI slope was great enough), then ring width might remain constant or even increase with time. If long-term ring width does not continue to decrease (as illustrated in Figures 1.6 and 1.7), then death of ancient trees is due to something other than failure to sustain a minimum ring width.

The classical ring-width trend approximates a die-away curve. Simulations suggest that, given enough time, the trend should come closer and closer to graphing as a horizontal line with no slope (Figure 6.9).

As a means of testing this, we can use linear regression to describe the linear BAI trend of real trees, project that into the future, and calculate the ring-width

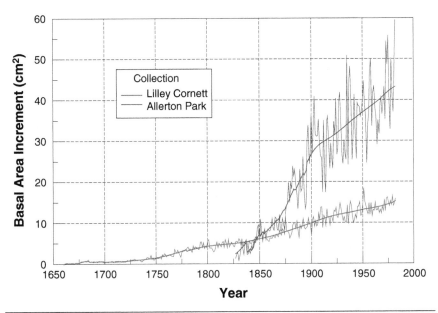

Figure 6.11   Raw BAI and BAI trends of two white oak collections. The faster-growing trees (blue) are from Allerton Park, IL, and the slower-growing trees (red) are from the Lilly Cornett collection in eastern KY (collected by Ed Cook). The initial 75 years of each collection appears to represent growth before the majority of the collection had achieved canopy status. Data were smoothed with a 60-year cubic spline to obtain the smoothed BAI trend of individual cores.

series that would be required to produce the projected BAI trend. We will illustrate this with data from a relatively young white oak collection from a good quality site at Allerton Park in central Illinois, and with data from an older, more slowly growing white oak collection from the Lilley Cornett Tract in eastern Kentucky (Figure 6.11).

Allerton Park White Oak

It was estimated that the Allerton Park collection attained canopy size circa 1900. This was based on both the median year when cores indicated a 10 cm radius and the shape of the BAI trend. Just as it is possible to use ring width to obtain the radius length needed to calculate the BAI (Eq. 6.1), so too was it necessary to include radius length in order to back calculate width (w) from area (BAI). In other words, depending on the diameter of the stem, different width series can produce the same BAI series.

Two ring-width simulations were run that were based on the BAI slope that was determined from regression. From the BAI trend data, ring widths were back calculated for the approximate number of years that the trees were in the canopy, 1900–1992, and then projected forward to the year 2200. This produced a decreasing ring-width trend (red curve, Figure 6.12). When the regression of the BAI trend of the canopy years was projected backwards, a value of 0 was reached for 1758 (Figure 6.13). Thus, if the BAI trend of the understory tree had been the same throughout its lifetime, the trees would have dated back to 1758. Back calculating ring widths from this scenario would have resulted in a constant ring width for years in the canopy (blue line, Figure 6.12). Assuming that cores of some of the oldest trees of the collection had come close to the center ring, it would appear that even the oldest trees may not have pre-dated 1825. Thus, it took no more than 75 years for the oldest tree to reach canopy size by 1900. By contrast, if the trees had grown at a constant BAI, it would have taken 140 years to reach canopy size. In other words, the collection trees grew at a greater rate as understory trees than they did after reaching the canopy in 1910.

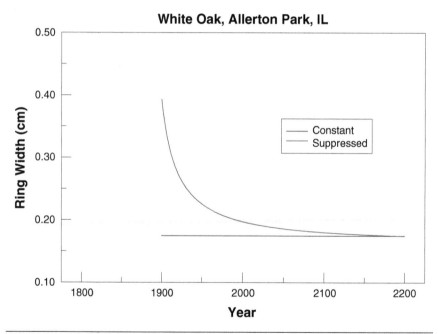

Figure 6.12 Simulated ring-width trends for canopy years (dbh > 20 cm) of Allerton Park white oak collection. Both simulations are based on the same BAI trend slope (0.19203, determined from regression for canopy years, 1900–1992). A decreasing ring-width trend (red curve) is the result of projecting ring width following an increase in competition between trees attaining canopy stature. Constant ring width (blue line) resulted when ring width was projected from a hypothetical situation in which the BAI trend had been the same slope throughout the life of the trees.

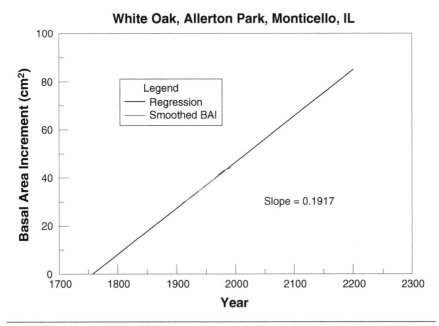

Figure 6.13   BAI trend of Allerton Park white oak collection. BAI (smoothed with a 60-year spline) is shown for years that radius ≥ 10 cm (1910–1992). Regression of the BAI was plotted from 1758 (year at which BAI = 0) to year 2200.

A greater growth rate meant that the pre-1900 BAI slope of the collection was greater than that after 1900. This suggests that the trees were established in a relatively open forest area allowing for a rapidly increasing crown size as the young trees grew from small tree, to sub-canopy, to canopy level. Then, after the canopy closed, BAI slope would have decreased as a result of increasing competition from surrounding trees. It appears that canopy closure and attainment of canopy size (10 cm radius) must have occurred at about the same time.

Nearby Monticello was laid out and named in 1837 (Piatt 1883). At the time of settlement, the native vegetation was tall-grass prairie, except for wooded areas along streams. The major stream of the county, the Sangamon River, flows through Allerton Park. The Allerton Park white oak collection was from an upland site adjacent to the Sangamon River. Except for an occasional burr oak grove, there just weren't any trees away from streams on the prairie. It would seem that the area that is now Allerton Park could have been a prime source in the 1830s for medium-sized trees from which to cut cabin logs. Some of the largest trees would likely have gone soon thereafter with the establishment of sawmills in the area. Removing only a part of the overstory would have been sufficient to allow rapid growth rates shown for the Allerton Park collection prior to the canopy again closing around 1900.

Lilley Cornett White Oak

The ring-width simulations were repeated using the Lilley Cornett collection (Figure 6.11). Diameter data suggest that canopy size was not attained until around 1845 for much of the collection, although there were a few trees that reached canopy size before 1750. Thirteen cores contained inside rings dating from 1660–1698. The dates at which these same cores reached a radius length of 10 cm ranged from 1786–1879.

The BAI trend of years since attaining canopy status was described by the regression of the BAI trend for the years 1845–1982 (Figure 6.14). Simulating ring width to the year 2200 shows a decreasing trend that tapers off as it approaches a near constant ring width (red curve, Figure 6.15). As with the Allerton Park trees, when growth rate in the canopy is projected back at the same rate and then summed, ring width remains at a constant (blue line, Figure 6.15).

The Constant Ring-Width Trend in Old Trees

The Allerton Park white oak collection might be considered as having trees of fairly typical ages for a secondary or tertiary forest in the eastern deciduous

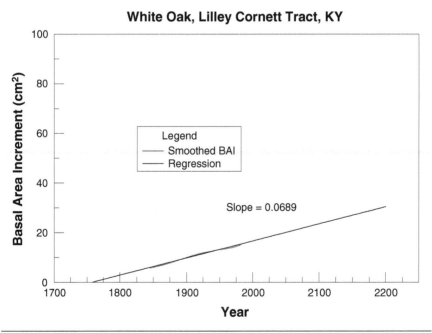

Figure 6.14 BAI trend of the Lilley Cornett Tract white oak collection and regression based on the BAI trend from 1845–1982.

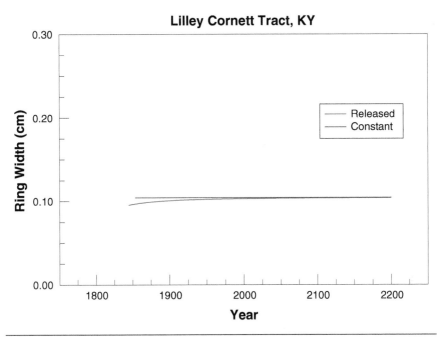

Figure 6.15 Simulated ring-width trends for canopy years (dbh > 20 cm) of Lilley Cornett white oak sample 103241. Both simulations are based on the same BAI trend slope (0.0806, determined from regression for canopy years, 1801–1982). An increasing ring-width trend (red curve) is the result of a decrease in the slope of the BAI trend when the tree attained canopy stature. Constant ring width (blue line) resulted when the ring width was projected from a hypothetical situation in which the BAI trend had been the same slope throughout the life of the tree.

forest. The simulated ring-width trend of the collection continued to decrease for 200 years after entering the canopy (Figure 6.12). This may be fairly typical of canopy trees of many forests in the East. In other words, we should expect that increment core samples from most deciduous forests in eastern North America will show a decreasing ring-width trend.

The Lilley Cornett collection from eastern Kentucky contains trees that are unusually old. There are, of course, other old sites in the East, though they are not common. The trend of increasing ring width illustrated for the collection would be the expected trend for trees that existed as suppressed understory trees before finally breaking into the forest canopy. Overstory trees from virgin or near-virgin forests would be expected to show this pattern. Most disturbed forests, such as the Allerton Park site, are too young to have trees displaying many decades of constant (steady state) ring-width trends.

In either case—decreasing width or increasing width—the long-term trend is toward a near constant ring width. A constant ring width necessarily means that amount of growth, expressed as BAI, continues to increase. Obviously, as will be discussed in the section *Killing a Tree by Size or Age*, a tree cannot continue a trend of increasing growth forever.

De-Trending and Time Stability

One of the two major purposes of calculating ring-width indices is to de-trend the series, thereby removing the non-climatic component from ring-width data. The second purpose, as described in Chapter 3, is to render the data more time stable. Time stability results because the amplitude of yearly variations in raw ring-width data are roughly proportional to width. Time stable data are very important in climatic reconstructions from tree rings.

BAI values and the amplitude of yearly variations become greater with time, whereas they become smaller with time in a ring-width series. This makes BAI, as well as ring width, a bit difficult to compare among trees of distinctly different size or growth rate. As discussed, after a tree attains canopy status, there is a tendency for the BAI trend to be positive and near-linear. Thus, an easy way to compare trees of differing size is to simply compare values of BAI slope.

The constant ring-width simulations for Allerton Park (0.1748 cm at a slope of 0.1920) and the Lilley Cornett Tract (0.1047 cm at a slope of 0.0689) suggest a relationship between slope and constant ring width. That relationship is graphed in Figure 6.16.

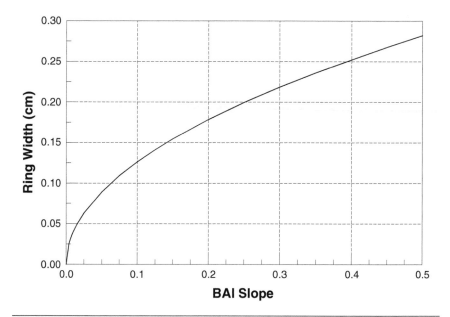

Figure 6.16   Hypothetical steady-state ring width by BAI slope.

The relationship graphed in Figure 6.16 allows determination of the steady-state ring width; that is, the ring width that will be approached in the long term for any given BAI slope. Whether ring width preceding this state is wider or more narrow than the steady state depends on whether the tree was in an open-canopy forest (decreasing ring width before steady state) or closed-canopy forest (increasing ring width before steady state).

PROJECTING GROWTH TRENDS FORWARD IN TIME

I have had foresters show me large, old trees that they said were *mature* and should be removed to allow growing space for smaller trees. Their basis for calling a tree mature was (1) it was large and had lost or was losing a large lower branch, and (2) the outer tree rings were quite small. Whether or not this description of mature is very commonly used, the long-term BAI trends suggest a different interpretation. Even if ring widths have reached a steady state, BAI may be continuing to increase. Perhaps removing a large tree that is still accumulating new material at an increasing rate might not always be the only way to go.

The foregoing discussion of long-term trends in ring width was based on projecting a linear BAI trend into the future and then back-calculating the ring-width trend. So long as a linear trend is expected to continue, a simple summation (accumulation) of the projected BAI values provides an estimate of accumulative growth for the period.

We can illustrate this using the BAI projections of the Allerton Park and Lilley Cornett collections that have been previously described. For example, the future 25 years of growth of the collection trees of the Allerton and Cornett collections can be obtained by projecting the BAI regressions forward 25 years and then summing the projected values (Table 6.4). On a per year basis, growth (expressed as BAI) of the projected years compared with the growth after reaching canopy size jumps from 16 to 20 cm^2 for Allerton and 11 to 16 cm^2 for Cornett. No matter how we look at it, these large trees are adding new growth at an increasing rate. Thus, it does not appear that we can use small ring widths of the big trees as an excuse for removing them.

In keeping with dendroclimatic standards, the Allerton Park and Cornett collections were made by selecting canopy dominants to represent the longest records with the least interference from crowding and competition among trees. If, on the other hand, trees had been selected to be representative of the stand, as is done in forest inventory sampling, then BA estimates and BAI trend determinations from growth projections could provide estimates of future growth for the stand, not just canopy dominants.

Table 6.4 Increase in BA for years in the canopy (from the year of entry into the canopy through the year of collection), and for 25 years beyond the collection year for the mean Allerton Park and the mean Cornett sample trees

	Allerton Park, IL	Cornett Tract, KY
Δ *BA*, years in canopy	1467 cm² (1900–1992)	1472 cm² (1845–1982)
Mean *BAI*, canopy years	16 cm²	11 cm²
Δ *BA*, 25-year projection	495 cm²	391 cm²
μ *BAI*, projected years	20 cm²	16 cm²

Impressive as the growth rates of these trees may be, we may not want to leave the big tree just because their growth rates have not slowed. In treating the forest stand as a crop, what difference does it make if the new growth is concentrated on a few large trees or scattered among several smaller trees? It may come down to stand dynamics. How is the long-term productivity of the stand affected by either leaving or removing the large trees? I'm not sure that we have all the answers to this yet. Perhaps some inroads to answers may be in combining some good tree-ring work with some innovative forest modeling.

ADDITIONAL FACTORS AFFECTING GROWTH TREND

The Photosynthate Potential

We can think of tree growth (tree rings) as an expression of surplus photosynthates. Food produced by the tree is used to supply energy to carry on various metabolic processes that serve to keep the tree alive. As the tree becomes larger, more and more energy is required just to move water from the roots to the leaves. In addition, energy is required to move food and growth regulators from the leafy portions of the tree to areas of physiological activity throughout the tree. If any energy is left over, it may be stored or utilized in the production of new tissues. This, of course, is greatly oversimplified, but the concept is a good one. On the other hand, regardless of the proportion of energy used for metabolic processes, the maximum amount of growth is limited by the size of the crown. Indeed, the shape of the BAI trend may be a reasonable representative of the history of the photosynthate potential of the crown.

While crown size is increasing, photosynthate potential may be simply a function of crown size. Each layer of leaves reflects some light, absorbs some light, and transmits some light. The percentage transmitted can then be utilized by the next lower leaf layer. This can continue to lower levels until there is not enough light to support photosynthesis. In younger, open-grown trees, a simple measure such as crown diameter may provide a good measure of photosynthate potential.

When a tree is undergoing some sort of growth decline, as indicated by a BAI trend curve (decreasing BAI slope), the crown thins and photosynthate potential decreases. During such a situation, the crown may have thinned even though the crown diameter may not have decreased. In this case, the BAI trend may still be reflective of photosynthate potential, though crown diameter is not.

The BAI time trend can often be expressed graphically as a straight line with a positive slope. Thus, the amount of growth is a bit greater each year than the year before. This is easy enough to understand in a young, usually pre-canopy-sized crown. We don't expect the crowns of forest-interior trees to become as large as those of open-grown trees. Indeed, we expect that in a closed-canopy forest, once a crown has gained canopy status, its diameter may change slowly or very little with time. At the same time, the BAI continues to show a positive trend, implying that photosynthate production is continuing to increase even though crown diameter may be constrained by surrounding trees.

It has already been mentioned that once a tree has established itself in the forest canopy, the length of the marketable bole does not increase appreciably. At the same time, tree height is expected to continue to increase. Those who have done any low-altitude flying over the Great Dismal Swamp (coastal plain, Virginia and North Carolina border) have commented that a number of baldcypress trees (*Taxodium distichum* (L.) Rich.) rise above the existing forest canopy. It can be seen from the air or from the ground that the tops are missing from these old giants. The trees were left at the time of original lumbering because they were not sound. It would seem possible that these old trees and hence, the original forest, could have originally been half again to nearly twice as tall as the top of the present canopy.

An effect of increasing height of a canopy tree with age may simply be to vertically elongate the crown. Elongating the crown enough to distend it from a spherical shape may serve to allow enough of an increase in the surface area of the crown to affect an increase in photosynthate production.

AGE OR SIZE FOR ENTRY INTO THE CANOPY

There are almost as many different ways to describe forests as there are authors to describe forest vegetation. Each author seems to have a unique concept of forest layers. The simplest concept has been to distinguish overstory and understory. Various authors may divide each of these two layers into additional layers.

Gareth Gilbert, working at the Neotoma research valley in the Hocking Hills of southeastern Ohio, distinguished three layers as canopy, sub-canopy, and small tree. Though these were distinguished on the basis of relative heights in the forest, he also identified them by dbh size classes:

Small-tree-size class	1–4 inches (2.5–10 cm)
Sub-canopy-size class	4–8 inches (10–20 cm)
Canopy-size class	≥ 8 inches (≥ 20 cm)

Neotoma was heavily cut over shortly before Ed Thomas bought the land in the early 1930s. Thus, when Gilbert did his work there in the 1950s and 1960s, the re-growth forest was still relatively young. Sometime after Gilbert's work, I did a rather informal survey of mixed age stands in numerous locations in the Midwest and mid-Atlantic regions. The intent was to estimate the dbh when the white oak trees entered the canopy. The approach was to ascertain the extremes in two groups: the greatest dbh of trees still in the sub-canopy layer, and the smallest dbh of trees in the canopy. Both groups seemed to be converging on 20 cm dbh. Why did I bother? This was the same result that Gilbert had gotten some two decades earlier working collectively with the canopy species at Neotoma.

A tree enters the forest canopy when it reaches about 20 cm dbh, regardless of age. In other words, the tree is ready to enter the canopy when it attains a certain physical size, regardless of whether it happens to be 20 or 200 years old at the time. In Chapter 2 it was shown that graphing by ring number from the center of the tree was a good method to determine shape of a curve. Unfortunately, that doesn't work too well with these old trees in which the trees have been in the understory differing amounts of time (Figure 6.17). When the graphs are lined up by the year the trees reached canopy size (Figure 6.18), the curves for growth since entering the canopy come closer to lining up.

KILLING A TREE BY SIZE OR AGE

We know that BAI continues to increase with age. As stated, this cannot continue forever. The fact that I have not seen BAI decrease, or even just level off, from age alone is rather curious. Sampling trees that have died recently enough that outside rings are still present might reveal how quickly healthy trees progress from a normal growth rate to death. I haven't exactly gone around looking for dead trees to sample, but in attempting to obtain long tree-ring records, one would, of course, be sampling the larger and often older trees of a given site. It might be expected that sooner or later, by chance alone, I would have run across and sampled trees that were starting to decline from nothing more than age.

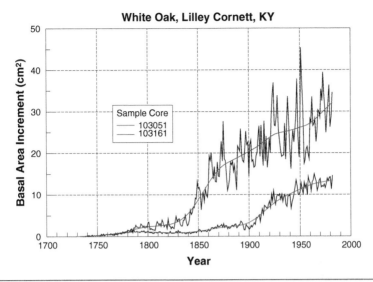

Figure 6.17   BAI trend of two white oak trees. The time at which each tree reached canopy size was assumed to be near the time of an increase in the amplitude of width variation. Tree 103051 (red) reached 10 cm of measured radius in 1923 and was estimated to have reached canopy size around 1900. Tree 103161 (blue) reached 10 cm of measured radius in 1856 and is estimated to have reached canopy size around 1850. Because the trees are essentially the same age, alignment by years is essentially the same as alignment by ring number (tree age).

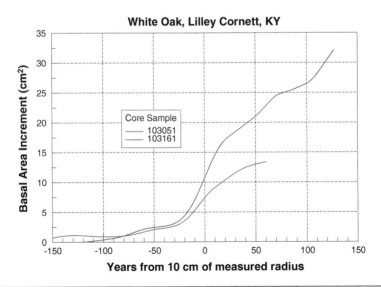

Figure 6.18   BAI trend of two white oaks aligned by the year that the measured radius of the sample core reached 10 cm. Compare with Figure 6.15.

On the other hand, perhaps the reason I haven't seen them is because they don't exist. In other words, it may be possible that when a tree dies of old age, it does it in just a few years; that is, so quickly that no decline or change in growth rate is apparent in the long-term trend of the tree-ring record.

What breaks down, allowing a tree to die of old age? A rather popular belief, or assumption, is that there is some sort of aging mechanism involved with the cambium. Perhaps, but I'm not convinced that this is the major part of the story. I think that it could just as easily be argued that the cambial activity at any given spot in the tree is basically just a function of food and water availability to the cambium at that point. For example, when a young tree forms adventitious sprouts more readily than an old one, is it because the cambium is younger or is it simply that the sprouts on the younger trees are closer to the food source?

Ring width has been described as an expression of cambial activity (Eq. 1.4). Ring-width simulations in this chapter showed that the long-term trend of ring width is toward constant ring width. Taken together, this suggests that cambial activity reaches a steady state in old trees, further suggesting that old trees do not die because cambial activity is on some sort of long-term decline related to age.

The foregoing discussion indicating trees enter the canopy by size, not age, would go along with an argument that old trees are killed by size, not age. The oldest tree that I ever sampled in the Washington, D.C. area was, in 1965, a 400-year-old chestnut oak (*Quercus prinus* L.) overlooking the old George Washington canal on the Virginia side of the Potomac River. The tree apparently had existed for a good many years in the understory before having the opportunity to move into the canopy. The tree was nearly 200 years old, but only about 25–30 cm dbh when the canal was built. At the time I sampled it, the tree looked better than, and was growing as well as, an equally large 250-year old tree of the same species a few meters away. That both trees had grown at about the same rate after entering the canopy would suggest that it made little difference that it took one tree 150 years longer to reach the canopy.

If a tree does not happen to be killed by disease or mechanical injury, and if the cambium of a large tree is potentially no different than the cambium of a small tree, what eventually kills the tree? It may be something as simple as there being a maximum potential size for a given species in a given habitat. Perhaps then, once a tree approaches that size, it is, in effect, in balance with its environment. Thus, once a tree reaches that point, any environmental aberration (such as a drought) that reduces photosynthate production or increases photosynthate demand, would be enough to tip the scales and the tree would die within a very few years.

Realistically, for the time being, contending that trees might die from size, not age, should be regarded as no more than speculation on my part until further work is done.

SELECTED REFERENCES

Phipps, R. L. (2005). "Some geometric constraints on ring-width trend." *Tree-Ring Research* 61(2): pp. 73–76.

Piatt, Emma C. (1883). *History of Piatt County.* Shepard and Johnston, Chicago, Illinois. Facsimile: Higginson Book Co., Salem, Massachusetts.

7

APPLICATIONS OF BASAL
AREA INCREMENT

As described earlier, basal area increment (BAI) and BAI slope are useful tools to quantitatively assess tree growth. This chapter describes two examples of the use of BAI in ecological studies examining growth trends. The first example uses tree-ring parameters in an exploratory investigation of loblolly pine (*Pinus taeda* L.). The second example is of an examination of long-term growth trends of white oak (*Quercus alba* L.) from 47 collections scattered from New York and Michigan to Alabama and Missouri.

BAI TREND FROM DBH OR FROM TREE RINGS

Diameter at breast height (dbh) is commonly used to calculate basal area (BA). If dbh measurements are repeated over several years, BAI may be calculated as the change in BA between measurements. The data may be adjusted for number of years between measurements if mean annual BAI is desired. In other words, with any of the methods involving dbh measurement, growth determinations go forward in time from the first measurement. Further, precision of any annual changes in growth trend are limited by the length of time between measurements.

Because tree-ring widths are annual increments of tree radius, ring-width data can be utilized to quantify growth in much the same manner as dbh data. Tree rings can be used to extend the growth record of a given age or size of class backwards in time—and area or volume growth can be determined for each annual ring. In contrast, BA calculated from dbh obtained from stand resurveying often does not resample any of the same trees, cannot distinguish individual years, and cannot provide information backwards in time. BAI is a function of both site quality and tree size. Because annual BAI increases with tree size (being most pronounced on better quality sites), working with graphic BAI slope may be helpful.

DISMAL SWAMP LOBLOLLY PINE STANDS

Much of the discussion regarding BAI has been with regard to hardwood species. The long-term expected BAI trend for years that a hardwood is in the canopy approximates a straight line with a positive slope. At the same time we have also stated that we really didn't know the expected long-term BAI trend for softwoods. Here we will examine tree-ring data of three loblolly pine stands within the Great Dismal Swamp National Wildlife Refuge (GDS NWR). This is a simple exploratory investigation intended to compare and contrast pine stands in a hydrologically modified swamp.

When we were first introduced to the GDS NWR, one of the sites to which the acting refuge manager, Mary Keith Garrett, took us was a stand of loblolly pines (*Pinus taeda* L.) that was bisected by Lynn Ditch just north of Middle Ditch. At that time, we did not know if we could even distinguish rings in trees growing in the swamp, let alone get any information from the rings. To us, this site became the Lynn-Middle Ditch collection site. We collected samples from the site at four or more different dates between 1972 and 1976. Early sampling at the Lynn-Middle Ditch site indicated distinguishable rings, suggesting that the rings could be measured and, perhaps, crossdated.

The dredge spoils from the construction of Lynn Ditch after World War II (WWII) were worked into a woods roadway on the east side of the ditch. This meant that the most accessible part of the Lynn-Middle Ditch loblolly stand was to the east of the ditch road rather than across the ditch to the west. Because Lynn Ditch runs more or less north-south, its construction impeded the general eastward movement of water across the area. Thus, some eastward flowing water was intercepted by the ditch, meaning that the water level on both sides of the ditch probably tended to be a bit lower after ditch construction. The pine trees were widely spaced and thus did not provide a continuous pine canopy, and were interspersed with other species. Having seen that young pines established after a fire typically formed dense, nearly pure stands, the rather scattered stand here suggested that the stand east of the ditch had long since self-thinned, whether more or less continuously or in distinct episodes.

In 1976 we collected cores from a loblolly stand along Hudnel Ditch that appeared to be in the process of self-thinning. Five years later we harvested and sectioned a tree from the site, and the following year we put together another loblolly collection, this time from the Dismaltown area.

We made five additional tree-ring collections in the swamp as part of the FORAST program. FORAST, an acronym for Forest Research of Anthropogenic Stress on Trees, was a research program headed by Sandy McLaughlin at Oak Ridge. We wanted a sixth collection from the swamp and so used previously collected cores from the Lynn-Middle Ditch site. Unfortunately, when we were dating and measuring the collections for FORAST, we were unable to crossdate the Lynn-Middle Ditch collection; thus, no ring-width data from Lynn-Middle Ditch were sent to Oak Ridge.

The initial attempts at dating the Lynn-Middle Ditch core samples were made by Diane Ireley of the United States Geological Survey (USGS) Laboratory of Tree-Ring Research (LTRR) as we were processing collections for FORAST. She couldn't initially get the cores to crossdate, but later she was able to date 55 of the cores collected at the Lynn-Middle Ditch site. The last collecting that we did at the site was in August 1976 from the east side of the stand (farthest from the roadway), which had not been previously sampled. We collected 32 cores of which 23 are included in the 55-core collection.

Ring Counts

A very simple way to compare tree-ring collections is by counting the rings (Figure 7.1). Because we had attempted to obtain core samples that were to the center of each tree, ring counts can be regarded as being within a few years of actual tree age. Considering ring counts to be approximate indicators of tree age, it would appear that the Dismaltown and Hudnel collections are from single-aged stands (Figure 7.1). The range in ages of the trees at the Lynn-Middle Ditch site, however,

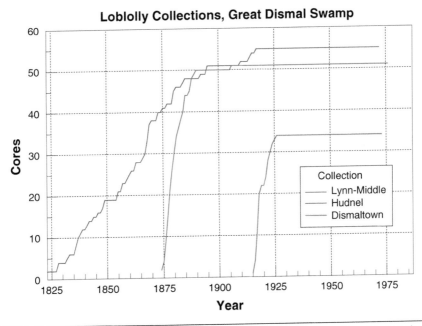

Figure 7.1   Number of cores by year for each loblolly collection. Considering the date of the inside ring as an indicator of tree age, note the great range in ages of the Lynn-Middle Ditch collection.

suggest a mixed-age stand. How the canopy at the Lynn-Middle Ditch site could have opened to permit establishment of new pine seedlings from time to time is certainly a matter of speculation. Although pine stands in the swamp appear to develop following a fire, no evidence of fire scars was noted on the trees or on increment core samples from the Lynn-Middle Ditch site. This suggests that fires have not occurred since the establishment of the oldest trees in the stand.

Ring Width

As a general rule, ring widths show trends that roughly follow a die-away curve from wide to narrower rings. Mean ring-width trends of the three loblolly pine collections (Figure 7.2) generally bear this out.

Trees from Dismaltown and Hudnel seem to be adding something to the story. Dismaltown cores show a sharp growth decline from the mid-1910s to 1930, followed by an increase from about 1945–1960, and then declining widths. The sharp decline is great enough to suggest increasing competition among young trees. A sharp increase in ring width often represents a release caused by removal of surrounding trees either naturally (self-thinning) or by lumbering. In this case, the smoothed ring-width increase is over perhaps 15 years. Interestingly,

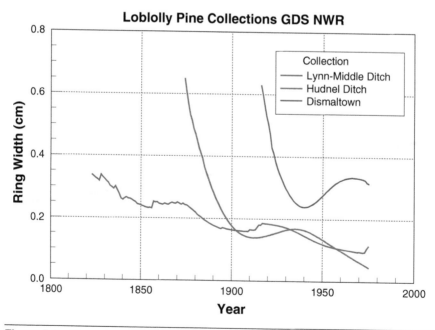

Figure 7.2   Mean ring-width trends of Lynn-Middle Ditch, Hudnel Ditch, and Dismaltown loblolly collections. Ring-width trends were obtained by smoothing ring width with a 60-year cubic spline.

nearby Washington Ditch was re-dredged just after WWII and, over a number of years, several other ditches were constructed. It may be that at least part of the ring-width increase (1945–1960) represents gradual root release resulting from lowered water levels following ditch construction.

The Hudnel stand appeared to be in the process of self-thinning at the time the core collection was taken. Ring widths preceding a self-thinning episode are expected to be relatively narrow. Other periods of narrow rings in the Hudnel Ditch collection occurred in the 1890s and early 1920s, and may also represent growth just before self-thinning.

The overall expected trend of decreasing ring widths is conspicuously interrupted by periods of increased ring width in all three collections. These apparent releases do not appear to have resulted from canopy openings caused by fire, but rather were more likely to represent releases following self-thinning.

BA

Ring area, that is, BAI, can be defined as an annual increment of BA. By adding (accumulating) BAI of successive years, BA for any and all rings can be determined; that is, tree rings allow us to look at BA backward in time and on a year-by-year basis. A graph of BA provides an indication of the growth history of stands or individual trees. The mean stand BA of the Dismaltown, Hudnel Ditch, and Lynn-Middle Ditch stands are presented in Figure 7.3. Although the

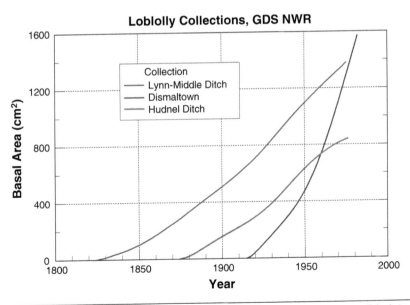

Figure 7.3 Mean *BA* by year of each of three loblolly pine collections. BA determined as accumulative smooth BAI.

Dismaltown collection has the youngest trees, their rapid growth has resulted in trees that by the 1970s were the largest in dbh of the three stands.

Ring-Width Indices

Identification of episodes of self-thinning may be discerned not only by decreasing growth rates, but also by growth releases. The curve fitting used to estimate the non-climatic component of growth obscures sudden shifts in growth such as might be caused by a release. A release is thus more easily identified from the climatic component, expressed as ring-width indices. Indices of the mixed-age Lynn-Middle Ditch collection suggest possible releases around 1867, 1903, and 1927 (Figure 7.4). Curiously, the 1903 possible release is also observed in the Hudnel collection. It seems unlikely that the two pine stands would experience natural thinning releases in the same year. A more plausible explanation is that the two stands are responding to an environmental factor common to both sites.

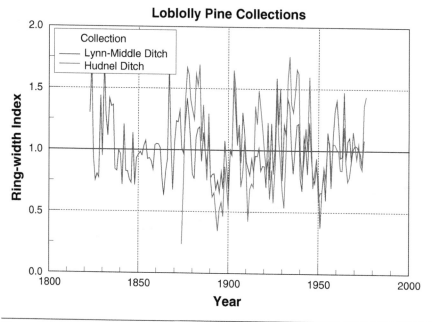

Figure 7.4   Mean collection ring-width indices of Lynn-Middle Ditch and Hudnel Ditch loblolly pine.

BAI

When examining collections, it is also important to look at data from individual trees. We may begin, however, by first looking at graphs of the full collections (Figure 7.5). The BAI curves contrast sharply with the ring-width curves (Figure 7.2) and, indeed, with the BA curves (Figure 7.3) for the same collections.

Both the Lynn-Middle Ditch and Hudnel Ditch collections show declining growth during the last 30 or more years. One question is whether this decline is related to suppression preceding self-thinning or to ditch drainage. That the BAI trend appears to be decreasing before ditch construction suggests suppression. Though not apparent in the smoothed BAI trend, raw basal area increment, BAI_R, indicates that the decline rebounded from about 1950 to after 1960 before again declining (Figure 7.6). The rapid rebound after 1950 may have been in response to root release after water table lowering from ditches. Perhaps continued lowering of the water table may have decreased to below the expanding root systems, thereby once again resulting in growth reduction. Thus, from BAI trends we might speculate that growth reduction prior to 1950 was from suppression, while growth reduction after 1960 may have been from a lowering water table.

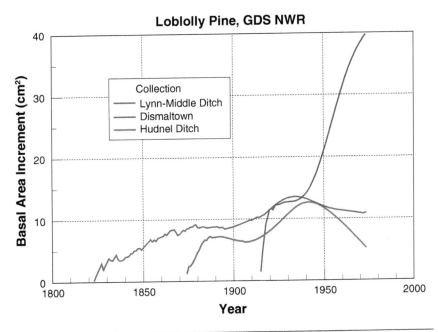

Figure 7.5 Mean BAI by year of three loblolly pine collections.

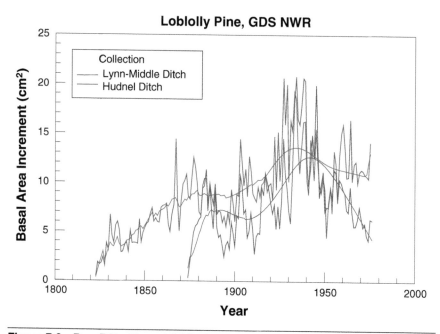

Figure 7.6   Raw BAI of Lynn-Middle Ditch and Hudnel Ditch loblolly collections.

An interesting feature of both the Hudnel Ditch and the Dismaltown collections is that BAI increases for the first few years and then levels off or decreases during the next several years. If, as with deciduous species, we regard the BAI trend of conifers as reflecting crown history, then the flattening of the BAI trend may suggest increased competition among similar-aged crowns in even-aged stands. Perhaps the reason we don't see this dip in the Lynn-Middle Ditch collection is because the BAI trend curve of the collection represents trees of many ages established at different times. Thus, we might not expect to see the suppression dip from similar-aged crowns at Lynn-Middle Ditch except in the oldest trees of the stand that may have been established in open conditions following a fire.

We can test this by graphing the BAI trends of some of the oldest trees of the Lynn-Middle Ditch collection (Figure 7.7). BAI trend curves showing a suppression dip in the Dismaltown and Hudnel Ditch collections was observed in 24 of the 34 cores and 44 of the 51 cores, respectively. As indicated in Figure 7.7, two of the 5 oldest trees showed a pronounced dip in the BAI trend. This suggests the possibility that growth suppression occurs as crowns of a young stand of loblolly pines begin to merge.

Because loblolly pine is relatively shade-intolerant, episodes of suppression and release are expected in this species growing in the GDS NWR. Indeed,

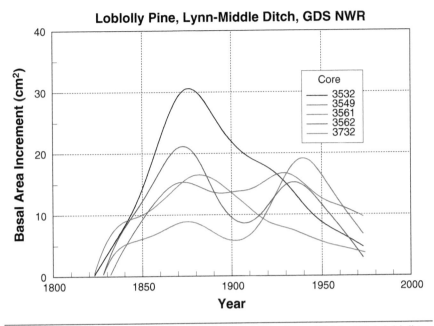

Figure 7.7 BAI trends for the five oldest trees of the Lynn-Middle Ditch loblolly pine collection.

suppression and release may be the primary factors controlling expected BAI trends. Death of some trees results in the release from suppression for many of those remaining. Loblolly pines at the Hudnel site were in the process of self-thinning at the time core samples were collected. The stand at the Lynn-Middle Ditch site was sufficiently older than the only remaining loblolly pines scattered throughout the stand. As the old pines died, opening of the stand favored the establishment of a few new pines although much of the open area became occupied by hardwoods, primarily red maple (*Acer rubrum* L.).

It appears that the long-term expected BAI trend of shade-intolerant loblolly pine in swampy sites is not quasi-linear, as it is for many hardwoods, but rather reflects cycles of stand self-thinning. Can we generalize that this could be a general model of growth trends for conifers? We only looked at three collections of one species in a limited area and habitat, but perhaps this was a start.

Three-Dimensional Growth

Transverse sections were taken at 6-foot (1.8 m) intervals along the trunk of a tree harvested from the Hudnel Ditch site. Peaks of the BAI trends of sections 3 through 11 (Figure 7.8) suggest that declining growth leading to self-thinning began in the late 1930s. Sections above 11 peaked at progressively later dates.

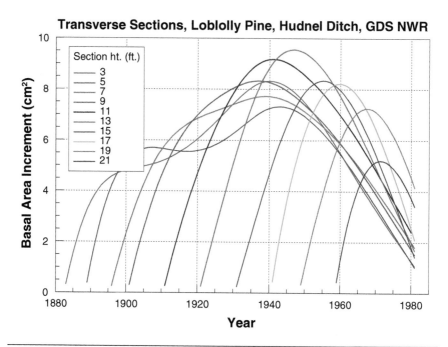

Figure 7.8   Mean BAI trends of transverse sections of a loblolly pine tree from the Hudnel collection site.

Just as BAI is expected to increase with successive radial growth at any given height, so it should increase basipetally within any given tree ring. This is not the situation in the sectioned tree after about 1940. Note, for example, that c. 1960, while growth of section 11 is decreasing, growth of section 19 is increasing. After c. 1967, the amount of growth per year of section 19 exceeds the amount of growth per year of all lower sections. Photosynthates seem to be allocated to the rings near the top of the tree at the expense of the same rings at lower heights.

Growth of inner rings of section 9 are increasing at the same time that growth of the same rings at lower sections have slowed a bit (Figure 7.8). For this particular tree, it appears that the period of increased allocation to the younger rings occurs for 10 or more years. This would imply that growth of the innermost 10 or so rings are not responding to climate (environmental variables) to nearly the same extent as older rings.

GROWTH TREND IN WHITE OAK

Work by the USGS LTRR in the 1980s with BAI trends was initially based on calculations from tree-ring measurements from hardwood increment core collections from Maryland and Virginia. Early on we concluded that the expected BAI trend of canopy-sized deciduous trees tended to be linear with a positive slope. At the same time, several of our collections were showing what appeared to be a negative departure (decreased BAI) from expected growth. Because we found that trees of different ages were all showing a synchronized decline starting in the 1950s, the decline did not appear to be age-related. By the early 1980s more and more evidence was accumulating to document increased air pollution in the mid-Atlantic region. Thus, the initial premise was that the decreased BAI slope indeed represented a growth decline, and that the decline was most likely a response to regional air pollution. The presumed decline was observed in about two-thirds of the collections that we examined. Nearly all of our collections were of a hardwood (Angiosperm) species. Our relatively few softwood (Gymnosperm) collections were insufficient to give us confidence concerning the expected growth trend in conifers.

Our study to examine the apparent growth decline was narrowed to a single species, white oak (*Quercus alba* L.). This choice was based primarily on the longevity of white oak and its wide geographic range. It was felt that older trees would provide long-term growth trends that could be used to identify growth declines prior to 1950. At the time of the study, there was concern regarding dieback in oak species, but less concerning members of the white oak group (subgenus *Lepidobalanus*) than of the red oak group (subgenus *Erythrobalanus*).

Determination of Growth Decline

Growth decline is a departure from expected growth. If the trees are being affected by air pollution, it might be anticipated that the rate of growth would decrease in a curvilinear fashion as environmental conditions deteriorated. The expected trend of ring-width data without an environmentally induced decline, however, is also curvilinear. Quantifying a curvilinear departure superimposed on a ring-width trend that is also curvilinear would be difficult to impossible. It was shown in previous chapters that a positive linear trend in BAI is a reasonable model for expected growth of typical deciduous trees. Departure from this linear trend thus is more easily identified and quantified.

One intent of calculating ring-width indices is to remove any growth trend, thereby rendering the data to be linear. However, the degree to which a growth decline may be expressed in indices will be dependent on the curve-fitting process used to generate the indices. As previously discussed in Chapter 3, a stiff curve applied to a ring-width series is expected to preserve essentially all of the climatic component in the indices, but will almost always include some of the non-climatic component. For this reason, it should be possible when working with indices produced with a stiff curve to distinguish at least some of the stronger growth declines. In other words, the stiffer the fitted curve, the more likely that some trend could remain in the indices. On the other hand, if a very flexible curve is used to define the non-climatic component, then little or none of the growth decline one wishes to examine may be preserved in the indices.

It may be possible to distinguish a growth decline from indices, but because indices are dimensionless they cannot be used to describe the severity of a decline in terms of actual growth quantity. We can do that with the BAI trend. However, regional air pollution and point source pollution are not the only causes of a decrease (or decline) in the BAI growth trend. It is entirely possible for growth decline to occur by natural causes. As pointed out in Chapter 6, the BAI trend appears to be a reasonable proxy of crown history, or more correctly, photosynthate production of the crown. An increase in competition among trees often results in a growth decline, though a decline is usually not as abrupt or as great as a growth increase caused by crown or root release. Because changes in competition in understory trees can be complex and can indicate natural cycles of decline and release, we wished to limit our investigation to canopy dominants, and to only those years since attainment of canopy stature. As explained in the section on *Age or Size for Entry into the Canopy* (Chapter 6), it was assumed that canopy size was achieved with a dbh ≥ 20 cm.

The original published white oak study was based on 89 collections. All were composed of trees that pre-dated 1900, though we were not able to include only trees large enough to have been in the canopy by 1900. Thus in the original study, some of the results were based on tree rings that had been formed before the trees reached canopy size.

Regional pollution in the United States was first noted in the northeastern region of the country. By the time elevated levels were being observed in the mid-Atlantic region, pollution in the northeast was starting to decrease and pollution was being documented in the Midwest. In other words, the magnitude and timing of the pollution varied by region. Further, rather than a curvilinear decline as might be expected from steadily worsening conditions, most of the collections in the original study showed declines that appeared to be linear. It was as if something (an event or events?) occurred between the late 1940s and early 1960s that caused a reduction in growth rate. From that initial study, it was concluded that the cause of the growth rate reduction was probably not regional air pollution.

Additional white oak collections have become available through the International Tree-Ring Data Bank (ITRDB), and thus it was possible to utilize data from 47 collections (of which 26 were from the 89 of the original study) that were composed of trees that were all in the canopy by 1900. The BAI trends (non-climatic component) of all 47 collections are included for reference in the Appendix. Because of the possibility that there might be regional differences in trend patterns, the 47 collections were arbitrarily divided into geographic regions by longitude: east (long. < 85°), middle (long. 85–90°), and west (long. > 90°).

The long-term trend in BAI, as discussed in Chapter 6, approximates a straight line with a positive slope. This model of the expected BAI trend at the time of the original study was based on having looked mainly at relatively young collections. Of concern was whether the linear trend continued in older trees. Obviously, a linear trend with a positive slope cannot always continue to increase. Long-term BAI trends of older trees have already been illustrated (Figure 4.9), though it might be helpful to illustrate them again with three older white oak collections: Lilley Cornett, Kentucky; Joyce Kilmer, North Carolina; and Mt. Lake, Virginia. Means of the first 10 cores of each collection to reach canopy status (20 cm dbh) are shown as smoothed basal area increment (BAI$_S$) trends in Figures 7.9, 7.10, and 7.11. Age information is summarized in Table 7.1.

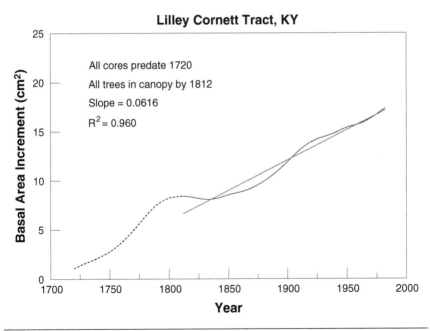

Figure 7.9   Long-term BAI$_s$ trend (and regression, red) of Lilley Cornett, KY, collection (collected by Ed Cook). Trend is the mean of the first 10 trees of the collection to reach 20 cm dbh.

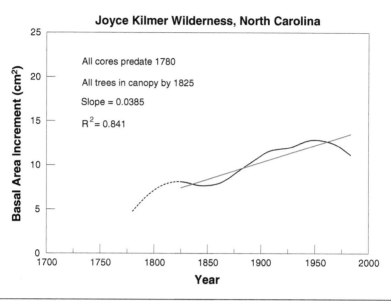

Figure 7.10 Long-term BAI$_s$ trend (and regression, red) of Joyce Kilmer, NC, collection (collected by Ed Cook). Trend is the mean of the first 10 trees of the collection to reach 20 cm dbh. Note the decline after 1950.

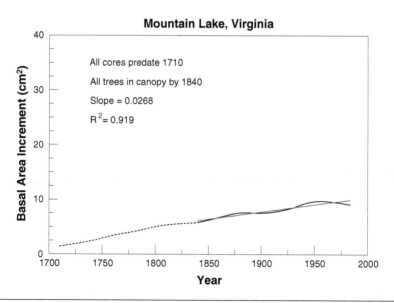

Figure 7.11 Long-term BAI$_s$ trend (and regression, red) of Mountain Lake, VA, collection (collected by Ed Cook). Trend is the mean of the first 10 trees of the collection to reach 20 cm dbh. Note the decline after 1950.

Table 7.1 Age information for the first 10 cores of trees to reach canopy size in the Lilley Cornett, Joyce Kilmer, and Mountain Lake white oak collections. [The trees were assumed to have reached canopy size at 20 cm dbh. Inside date is the date of the innermost measured ring.]

	Lilley Cornett	Joyce Kilmer	Mountain Lake
Inside date			
Oldest tree	1670	1641	1552
Youngest tree	1720	1780	1710
Date at 20 cm dbh			
First tree	1773	1768	1686
Last tree	1812	1825	1840
Age at 20 cm dbh	110 ± 14	89 ± 32	128 ± 14
Age in 1950	264 ± 18	253 ± 35	284 ± 52

Deviations in BAIs from the linear trend are apparent in each collection. Whether any of these deviations is due to regional factors expected to alter growth can only be ascertained by examining a number of collections from the region. With regard to the three collections illustrated, for example, none of the trend deviations is time synchronous. From this one might infer that in a region large enough to include all three collection sites, none of the deviations appears to be a response to phenomena in common among the collections.

Each of the 47 collections utilized for the study was composed of two increment cores from each of a dozen or so canopy white oak (*Quercus alba* L.) trees at a given site. Samples that had not attained a 10 cm radius length by 1900 were deleted from analysis. Total radius length of each core (needed to calculate BAI) was measured directly from each core of our own collections. For other collections, the assumption was made that measured radius length was not appreciably different than total radius; thus, the sum of all ring-width measurements was used to represent total radius length.

The smoothed BAI trend was calculated as described in previous chapters. The ring-width series of each core was smoothed with a 60-year cubic spline to obtain an estimate of the non-climatic width trend. The resulting smoothed data of the individual cores were then averaged to obtain a smoothed BAI trend for each collection. The smoothed BAI collection trend represents the mean growth trend (non-climatic component) of the dominant canopy trees of the stand sampled.

Our original study describing a growth decline after 1950 had suggested a link between site quality and the BAI slope of the pre-1950 BAI trend. From that original study, a BAI slope between 0.1 and 0.2 was arbitrarily chosen to represent moderate site quality. A slope greater than 0.2 was considered to represent good site quality and a slope less than 0.1 was considered to be poor site quality. These arbitrary settings resulted in the original 89 collections being divided into about an equal numbers of collections classified as good, moderate, and poor.

The original data set of 89 collections included 77 collections from the FORAST program. Many of the FORAST collections were made by cooperators with biological or forestry training but with little or no previous experience in dendrochronology. Thus, trees within sample sites were often selected randomly. The result was that collection strategies may have favored canopy dominants, but not necessarily individual trees yielding ring patterns showing particularly good climatic sensitivity. The 47 collections utilized for the re-examination described here were made mainly by dendrochronologists who emphasized record length and climatic sensitivity. Greater climatic sensitivity is usually expected from sites that are particularly limiting to growth. For the re-evaluation study, the good site quality class was eliminated and the boundary between the moderate and poor classes was adjusted to end up with a nearly equal number of collections in each class. The moderate sites were distinguished as having a graphed initial BAI slope value ≥ 0.09, and the poorer sites were those with an initial BAI slope value < 0.09. Following this classification adjustment, each of the two site quality classes (moderate and poor) was represented by at least five collections from each of the three regions, western, middle, and eastern (Table 7.2).

Table 7.2 Number of white oak collections by site quality within states of each geographic region. [Site quality based on initial BAI_S of post-1900 collection regression as indicated in the Appendix: better = ≥ 0.09 and poor = < 0.09. See text for explanation.]

	Moderate	Poor	Total
Eastern Region (< 85° W Long.)			
Kentucky (eastern)	0	1	1
Maryland	1	0	1
Michigan	0	1	1
New Jersey	1	0	1
New York	0	1	1
North Carolina	0	2	2

Continued

	Moderate	Poor	Total
Ohio	3	0	3
Pennsylvania	2	0	2
Virginia	1	3	4
Totals	8	8	16
Mean initial BAI slope	0.196	0.045	0.120
Middle Region (85–90° W Long.)			
Alabama	1	0	1
Illinois	6	3	9
Indiana	1	0	1
Iowa (eastern)	1	0	1
Kentucky (western)	1	0	1
Missouri (eastern)	0	2	2
Totals	10	5	15
Mean initial BAI slope	0.160	0.041	0.120
Western Region (> 90° W Long.)			
Iowa (western)	5	8	13
Missouri (western)	1	2	3
Totals	6	10	16
Mean initial BAI slope	0.131	0.048	0.079
Grand Totals	24	23	47
Mean initial BAI slope	0.164	0.045	0.106

Deviation in Long-Term Trends

Growth decline may be quite simply defined as a change in the expected trend. Although the BAI trends approximate linear trends, none is a perfectly straight line. An objective of the study was to examine changes, positive as well as negative, to the linear trend model. Two aspects of change were considered. One was the change in BAI slope from one linear segment to the next. A change was arbitrarily considered as appreciable if the change in slope was > 0.1. The other aspect regarded the abruptness with which the change in slope occurred. A very gradual change over several years was not considered appreciable even if the slope change was ≥ 0.1. Slope change was considered appreciably abrupt if any of the second differences of smoothed BAI values was ≥ 0.01. Length of the change segments was determined by the years in which the second difference was ≥ 0.01.

The Initial Slope

The understory of a closed forest is expected over time to be subjected to varying degrees of shading and crowding. This means that BAI_S (non-climatic) trends of trees while they were in the understory contain information regarding understory stand dynamics. For this study, though, we were interested in growth unrelated to stand dynamics, so growth of understory trees was not considered. Because the expected trend of individuals in the canopy may be graphed as a straight line, the initial BAI slope was defined as being the first linear trend after the tree is well established in the canopy. Some trend changes associated with entry into the canopy are discussed in the next paragraph. Initial slope data of the 47 collections are presented in Table 7.3 and changes after the initial slope are summarized in Table 7.4. In the majority of collections, the initial slope was the first linear segment after all trees reached canopy status (assumed to be 20 cm dbh). Slope was determined by regression of linear segments of the graphed BAI data (see Appendix). Changes in slope are expressed graphically in Figures 7.12, 7.13, and 7.14.

Table 7.3 Basic information of 47 white oak collections selected for analysis. [Collectors: Ed Cook[1], Dan Duvick[2], Eric Jamison[3], Bob Long[4], Rubino/McCarthy[5], Dick Phipps[6], Dave Stahle[7], John Whiton[8]. Whiton, Jamison, and Phipps were with the USGS LTRR. *MS* = mean sensitivity of the selected cores. *IS* = initial slope. Selected cores were only those judged to be from trees in the canopy by 1900 (radius ≥ 10 cm). IS is the first positive linear BAI segment after all trees had attained 20 cm dbh. IS < 0.090, regarded as poor-quality sites, are shown in red.]

Collection	Coordinates	Cores Total	Cores Selected	*MS*	*IS*
Eastern Region (Longitude < 85° W)					
Alan Seeger, PA[4]		69	51	0.115	0.121
Andrew Johnson Woods, OH[1]	40N 81W	40	40	0.194	0.158
Cook Forest, PA[1]	41N 79W	25	13	0.153	0.145
Cranbrook Institute, MI[1]	42N 83W	24	24	0.155	0.073
Dark Hollow Trail, NY[1]	41N 74W	22	16	0.156	−0.001
Davis Purdue/Glen Helen, OH[1]	39N 84W	23	22	0.158	0.165
Dysart Woods, OH[5]	39N 81W	10	10	0.225	0.327
Hutchenson Forest, NJ[1]	40N 74W	36	36	0.174	0.121
Joyce Kilmer Wilderness, NC[1]	35N 83W	30	25	0.116	0.022
Lilley Cornett Tract, KY[1]	37N 83W	40	32	0.156	0.047

Continued

Collection	Coordinates	Cores		MS	IS
		Total	Selected		
Limestone Branch, VA[8]	39N 77W	32	32	0.145	0.368
Linville Gorge, NC[1]	35N 81N	42	19	0.158	0.051
Mountain Lake, VA[1]	37N 80W	26	20	0.162	0.043
Patty's Oaks, Blue Ridge, VA[1]	37N 79W	23	21	0.131	0.057
Stony Man, VA[3]	39N 78W	32	16	0.142	0.070
Yell Farm, MD[3]		34	30	0.139	0.159
Middle region (Longitude 85–90° W)					
Allerton Park, IL[6]	40N 89W	23	16	0.245	0.121
Babler State Park, MO[2]	38N 90W	37	37	0.184	0.007
Cameron Woods, IA[2]	41N 90W	24	9	0.212	0.236
Ferne Clyffe State Park, IL[2]	37N 88W	47	25	0.182	0.148
Fox Ridge State Park, IL[2]	39N 88W	34	24	0.203	0.101
Giant City State Park, IL[2]	37N 89W	52	46	0.173	0.078
Kankakee River State Park, IL[2]	41N 88W	30	18	0.144	0.113
Kickapoo State Park, IL[2]	40N 87W	21	10	0.200	0.092
Lower Rock Creek, MO[2]	37N 90W	34	16	0.168	0.066
Mammoth Cave, KY[1]	37N 86W	42	33	0.149	0.206
Lincoln's New Salem State Park, IL[2]	39N 89W	58	42	0.209	0.040
Pulaski Woods, IN[1]	41N 86W	22	22	0.177	0.193
Sandwich, IL[2]	41N 88W	18	12	0.177	0.314
Sipsey Wilderness, AL[1]	34N 87W	32	28	0.158	0.099
Starved Rock State Park, IL[2]	41N 89W	97	85	0.175	0.014
Western Region (Longitude > 90° W)					
Backbone State Park, IA[2]	42N 91W	72	11	0.182	0.000
Current River Natural Area, MO[2]	37N 91W	97	41	0.193	0.104
Dolliver Memorial State Park, IA[2]	42N 94W	30	14	0.176	0.083
Duvick Backwoods, IA[2]	41N 93W	88	17	0.222	0.294
Geode State Park, IA[2]	40N 91W	33	25	0.199	0.058
Lacey-Keosauqua State Park, IA[2]	40N 91W	27	12	0.209	0.005
Lake Ahquabi State Park, IA[2]	41N 93W	62	44	0.189	0.091
Ledges State Park, IA[2]	42N 93W	125	65	0.183	0.042
Merritt State Forest Preserve, IA[2]	42N 91W	33	27	0.204	0.099

Continued

Collection	Cores				
	Coordinates	Total	Selected	*MS*	*IS*
Pammel State Park, IA[2]	41N 94W	119	81	0.184	0.086
Roaring River, MO[7]	36N 93W	33	19	0.143	0.035
Saylorville Dam, IA[2]	41N 93W	73	23	0.172	0.098
Wegener Woods, MO[7]	38N 91W	39	29	0.209	0.079
White Pine Hollow State Preserve, IA[2]	42N 91W	16	16	0.184	0.100
Woodman Hollow State Preserve, IA[2]	42N 94W	94	57	0.165	0.019
Yellow River State Park, IA[2]	43N 91W	24	11	0.189	0.073

Table 7.4 Number of collections (out of 47) showing changed BAI slope > 0.1 by decade after the initial linear canopy segment. [Changed slope from the initial slope (see text) by decade in which changed slope began. + = increase > 0.1; – = decrease > 0.1.]

	1920s		1930s		1940s		1950s		1960s		1970s		Totals	
	+	–	+	–	+	–	+	–	+	–	+	–	+	–
E								4	2	1			1	6
M							1	1	2				1	3
W			3		1			1						5
Σ			3		1		1	6	4	1			2	14

Trend upon Entering the Canopy

Accession to the canopy was assumed to have occurred when a tree reached a dbh of 20 cm. When the canopy has been opened by lumbering or natural disaster, understory trees are released (expressed as increased slope of the BAI trend) until the canopy again closes. As the canopy closes, competition increases sharply enough that growth slows and is no longer in the *released* phase displayed after disturbance. Indeed, in some collections, particularly in the western stands, canopy closure appears to result in a period of suppression even after trees have attained 20 cm dbh. In other words, although 20 cm dbh was used as the size at which trees reached the canopy, attainment of canopy size may not have occurred in some of the western stands until the majority of the trees were even larger than 20 cm dbh (Figures 7.12–7.14).

Many understory trees in a closed forest display suppressed growth (BAI trend at low slope angle). When these trees eventually reach the canopy, they may be referred to as being released into the canopy, meaning that the growth

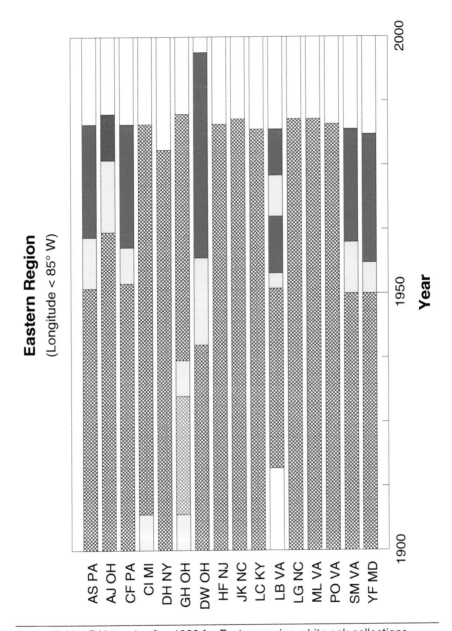

Figure 7.12   BAI trends after 1900 for Eastern region white oak collections. Black cross-hatch: initial linear trend after attainment of canopy size (dbh > 20 cm). Red cross-hatch: suppression (decrease) thought to be associated with canopy closure. Yellow: period of slope change between linear trends. Red: linear trend that is an appreciable decrease from previous (initial) trend. Blue: linear trend that is an appreciable increase from the previous (initial) trend. See text for further explanation.

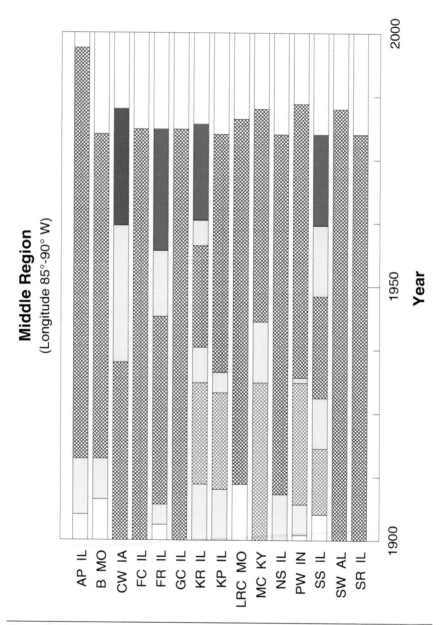

Figure 7.13 BAI trends after 1900 for Middle region white oak collections. Black cross-hatch: initial linear trend after attainment of canopy size (dbh > 20 cm). Red cross-hatch: suppression (decrease) thought to be associated with canopy closure. Yellow: period of slope change between linear trends. Red: linear trend that is an appreciable decrease from previous (initial) trend. Blue: linear trend that is an appreciable increase from the previous (initial) trend. See text for further explanation.

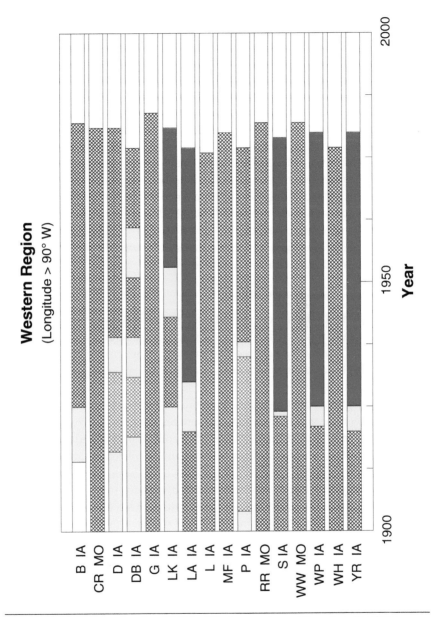

Figure 7.14   BAI trends after 1900 for Western region white oak collections. Black cross-hatch: initial linear trend after attainment of canopy size (dbh > 20 cm). Red cross-hatch: suppression (decrease) thought to be associated with canopy closure. Yellow: period of slope change between linear trends. Red: linear trend that is appreciable decrease from previous (initial) trend. See text for further explanation.

rate increases (BAI slope increases) as the tree crowns gain access to direct solar radiation. As pointed out in the previous chapter, this type of growth pattern is most common in old, established forests.

Nine of the 47 collections displayed a robust trend change that was inferred to be suppression associated with canopy closure. The slope change was > 0.1 at the end, and usually at the beginning, of the growth reduction. The reductions are identified in Figures 7.12–7.14 as red cross-hatch. These segments appear in bar graphs of collections 11, 12, 13, 22, 23, 31, 35, 37, and 40 (Figures 7.12–7.14). Examining the smoothed BAI trends of the collections (Appendix) suggests that in addition to the nine collections identified, other collections may also show the pattern though the trend change at the beginning and end of the reduction phase was not appreciable (< 0.1). The most obvious examples of these include curves for collections 17, 20, 34, 41, and 42. What is here inferred to be suppression associated with canopy closure is what we referred to in the original, published study as an unexplained Midwest pattern.

Trend of Decreasing Growth Rate

This re-examination uses more rigorous criteria for distinguishing appreciable slope change than was used in the original study. In the re-examination, 31 of 47 collections did not show an appreciable change in slope. Again, an appreciable change in slope was defined as a slope change > 0.1 along with an abruptness of change identified as a second difference > 0.01. Those showing no change were about equally distributed among regions. This may be comparable to the original study if the good site collections were removed from the original study.

As shown in graphs in the Appendix, two collections showed a distinct increase in growth (increase in slope > 0.1), namely collections three (Andrew Johnson, Ohio) and six (Cameron Woods, Iowa). In both cases the trees were growing rapidly, and it would not be too much of a stretch to speculate that there was a slight growth reduction that may have resulted with accession to the canopy. Thus, the growth increases may have actually been the initial trends after accession to the canopy. If this were the case, then there was no collection showing an increase over the initial trend. In any event, it does not appear that the increased trends are appreciable or in response to some regional phenomenon.

A growth reduction (or decline) was defined as an appreciable decrease (> 0.1) in slope after the initial linear trend. Using this criterion, 14 out of the 47 collections showed growth declines. Six were from the east region, three from the middle region, and the remaining five were from the western region (Table 7.5). All nine of the eastern and middle region collections and one of the western collections began a linear growth decrease in the 1950s or early 1960s. The remaining four western collections show decreasing growth beginning in

Table 7.5 Number of collections showing appreciable change in slope of the BAI trend. [An appreciable change in slope was considered as a change > 0.1 in which the change occurred abruptly enough that the second difference of the yearly slope values was > 0.01.]

Region	No Change	Decrease	Increase	Total
East	9	6	1	16
Middle	11	3	1	15
West	11	5	0	16
Total	31	14	2	47

the 1920s. These can be thought of as composing two groups; one showing decreased growth beginning in the 1950s (six collections in the eastern region, three in the middle region, and one in the western region) and a group of four collections from Iowa that all began declining in the 1920s. Thus, decreases starting in the 1950s are strongly skewed to the east.

Because of the intricacies of the curve fitting involved in creating BAI trends, it is probable that actual slope change did not occur quite as early as the beginning of the change period or quite as late as the end of the period of change. To say that most of the declines in growth began in the 1950s is probably about as precisely as it can be dated. About two-thirds of the collections in the original study showed growth declines starting in the 1950s or early 1960s. That study also showed a tendency for collections with the greatest growth rate to show the sharpest subsequent decline. Using more rigorous criteria in the present study and fewer faster-growing trees, 10 out of 47 (or about one-fifth) of the collections show a decreasing growth trend since the 1950s.

The original study, using less rigorous, subjective criteria, found similar proportions of declining growth trends in all regions. Regional air pollution is generally regarded to have first reached serious proportions in the Northeast and later in the Midwest and mid-Atlantic regions. Approximately equal proportions of collections in the original study showed declines across the regions at about the same time. It was concluded that regional pollution was not a likely cause for the noted declines. The new results shown here, in which declines starting in the 1950s are skewed toward the East, suggest that regional pollution, though still unlikely, cannot be ruled out as a possible cause for the decline.

Air pollution progressed gradually. Intuitively, worsening environmental conditions would be expected to result in a downward curving growth trend. Conversely, a decreased trend (or decline) that is linear could be the result of (1) an event that caused a permanent reduction in growth rate or (2) a permanent change in environmental conditions. The original study concluded that in most of the collections showing a decline, the decline was linear. This was used as an additional factor supporting the contention that the decline did not appear to be the result of worsening environmental conditions brought on by regional air pollution. The majority of collections used in the newer study and which show a decline seem to be linear. But, it is not at all clear cut. With reference to the collections (see Appendix for BAI trends of individual collections), good examples of linear declines were seen in the Fox Ridge and Yell Farm collections. Good examples of linear declines that were not designated as appreciable may be seen in the BAI trends of the Dark Hollow Trail and Patty's Oaks collections. The Cook Forest Fire Tower collection is an example of an appreciable decline that is curvilinear. Non-appreciable curvilinear declines may be seen in the Current River and Ferne Clyffe collections.

The four Iowa collections showing a growth decrease beginning in the 1920s or 30s are interesting if not a bit perplexing. It is as if growth began a decline with increased competition when the canopy started to close, but then recovery never occurred. The linear decrease of three of the four collections (all except Lake Ahquabi) seem to be bottoming out by the time the collections were made in the mid-1970s. It might be speculated that the collection locations, which are all near the western limit of the deciduous forest, may have been relatively open grown. Thus, when the trees reached 20 cm dbh they, by definition, had reached canopy size even though the canopy may not have yet closed. Then the decline starting in the 1920s may represent nothing more than increasing competition during a long, gradual closure of the canopy.

Interpretation of the BAI trend is not at all cut-and-dried. The technique is new enough that we really don't have all the answers. So, the bottom line has to be that based on our present understanding of the BAI trend, we have identified a trend that negatively deviates from the expected trend and is thus referred to as a growth decline. The cause of what appears to be a decline remains speculation. It has now been more than 30 years since the data collection of the original study. If the decline was the result of an event, then the decline might still be in progress. Is it? It was pointed out in the previous chapter that the natural death of trees appears to be controlled by size, not age. Decreasing the growth rate regardless of cause, might serve to increase the life expectancy of the tree as long as maximum potential size has not changed. On the other hand, if whatever caused the decline also decreased maximum potential size, then life expectancy may also decrease. This indeed sounds as if we are still in the beginning phases of really understanding what BAI trends of our trees can tell us.

We once heard someone say that he was not interested in pursuing botany as a career because, scientifically, everything had already been done. Thus, he said, he would have difficulty picking a research topic that had not already been researched by someone else. Seriously?

SELECTED REFERENCES

Phipps, R. L. and J. C. Whiton. (1988). "Decline in long-term trends of white oak." *Canadian Journal of Forest Research.* 18: pp. 24–32.

8

THE CLIMATIC COMPONENT: CORRELATION WITH ENVIRONMENTAL FACTORS

Entire books have been written about reconstructing climate from tree rings. Two classics are: *Tree Rings and Climate*, written by H. C. Fritts (1976) and *Methods of Dendrochronology*, assembled and edited by E. R. Cook and L. A. Kairiukstis (1990). A wealth of journal papers has followed. Most of these use ring-width indices as basic tree-ring data. Considerable effort has been expended in previous chapters to describe basal area increment (BAI) and to show its utility in ecological applications. Offhand, if one is interested in working with growth quantity, then it would make sense to use BAI as the parameter of growth when examining correlations with environmental factors. The BAI time trend has been described in previous chapters as tending to be linear with a positive slope; that is, to increase with time. A correlation of BAI with an environmental factor such as precipitation that does not increase with time will result in the most recent BAI values carrying the most weight. Thus, the time trend introduces a bias in the data. It would be necessary to remove the time trend from the BAI data in order to remove the bias.

Perhaps the easiest way to remove the time trend from a BAI series is to scale the raw data to the trend; that is, to divide raw BAI by the non-climatic trend (a smooth curve fitted to the data). The result, of course, is an index. If the curve-fitting process is equivalent to that used in calculating ring-width indices, then the BAI indices and ring-width indices should be equivalent. Thus, with regard to correlations with environmental factors, ring-width indices remain the parameter of choice simply because there is no advantage in converting to BAI before calculating indices.

The intent here is to present a general introduction to methods of extracting environmental information from tree rings, placing emphasis on a few aspects that may be of particular value in ecological studies. We follow with an example in which tree rings are used to reconstruct lake level data.

THE BASICS

As a general rule, an environmental factor correlates with growth inasmuch as it, or something that is in turn correlated with it, limits growth. If a shortage of water, expressed as rainfall, is limiting to growth, then as rainfall increases, growth increases, and as rainfall decreases, growth decreases. On the other hand, if water does not limit growth, then changes in the amount of rainfall have essentially no effect on changes in the amount of growth. One can imagine a well-drained site in a rain forest in which, even on a very dry year, there is always an ample supply of water. Whether rainfall is above or below normal might have very little effect on the amount of growth, resulting in little or no correlation between growth and rainfall. On the other hand, too much water under some circumstances may be limiting to growth in much the same way that too little water under other circumstances may be limiting.

If one wishes to examine relationships between a particular environmental factor and tree growth, then it is important to collect tree-ring material from sites in which that factor is limiting to growth. For example, consider hemlock (*Tsuga canadensis* (L.) Carr.) growing in New York and Virginia. In New York, hemlock may be found to grow in a range of hydrologic conditions; hence, it is not difficult to select hemlock sampling sites where water can be limiting to growth and from which good correlations between tree rings and precipitation might be possible. By contrast, Virginia is far enough south that hemlock is generally restricted to protected, steep, north-facing slopes that are continuously moist; that is, it grows in very restricted hydrologic conditions. Because such sites are expected to show little variation in moisture availability, correlations with precipitation may be quite weak. On the other hand, summer temperatures in Virginia can be great enough to limit hemlock growth. Thus, in very broad generalities, rainfall may be much more strongly correlated with growth of hemlock in New York than in Virginia, whereas temperature may be more strongly correlated with growth of hemlock in Virginia than in New York.

Climatic Sensitivity

Measured values of environmental factors of any given site in the deciduous forest tend to vary considerably from year to year. If tree rings from trees on the site do not show a correspondingly variable ring-width pattern, then correlations between growth and environmental factors at that site may not be particularly strong. Generally, the greater the year-to-year variability in ring width, the greater the chances that ring growth is responding to something in its environment, often related to climate. The mean variability of the rings is commonly referred to as mean sensitivity, *MS*, and is calculated from ring widths, *w*, as:

$$MS = \frac{1}{n-1} \sum \left| \frac{2\left(w_n - w_{n-1}\right)}{w_n + w_{n-1}} \right|$$

As explained in foregoing chapters, *MS* is the mean of the annual sensitivity (*AS*) values for each ring in a series. The *AS* values are calculated as the change from the previous ring relative to the mean of the present and previous rings. To calculate the collection mean, *AS* is expressed in absolute terms.

The term *climatic sensitivity* is a generalized term that is usually used to imply correlation with climate. A climatically sensitive collection is one that is inferred to be, or has actually been shown to be, correlated with climatic factors. Generally, the greater the *MS* of a series, the greater are the chances that the series will be climatically sensitive; that is, it will correlate well with climate. Informally, the terms *MS* and *climatic sensitivity* are sometimes used interchangeably.

The purpose of dendroclimatological studies is to produce climatic reconstructions that contain as much climatic information as possible. In a given region this translates to carefully selecting species and sites, and then carefully selecting individual trees. Once a tree is selected, then the sides of the tree to sample are selected. Unfortunately, many ecological studies do not have the luxury of a broad range of selections. More often than not, one must work with whatever trees are at hand, whether or not they contain rings that are particularly climatically sensitive. Suppose, for example, that one wishes to examine how precipitation and temperature limit the growth of gap-edge trees versus non-gap-edge trees at a particular site. Only limited selection would be possible, such as distinguishing crowding and dominance classes.

Water Availability

Throughout temperate regions of the world, water availability is probably the single most important factor affecting tree growth. A great variety of factors, such as edaphic conditions, slope exposure, and bedrock type may each correlate with growth due to their effects upon water availability. We are presently experiencing a period of global warming. The extent to which warming affects tree growth of trees of a particular location may be dependent upon the degree to which water availability to those trees is affected.

Slopes may be described in terms of surface and soil water collection or dispersal. A convex slope, such as a ridge or nose, is an area of water divergence. Water that falls on a convex slope tends to flow away from the convex slope. A concave slope, such as a cove or base of a slope, is an area of water convergence.

Water will collect in a concave slope from upslope in addition to that which fell directly as precipitation. A straight side slope that is neither convex nor concave would be expected to lose about the same amount of water that it gains, such that soil water inflow from higher elevations and soil water outflow to lower elevations are about equal. In conversation, the late John Goodlett at Johns Hopkins used to refer informally to these three slope types as *nose-aceous* (convex slope), *cove-aceous* (concave slope), and *side hill-aceous* (straight slope).

One might expect that an area of water divergence (convex slope) would hold the most promise from which to obtain a climatically sensitive tree-ring collection simply because water at this site would be in the shortest supply and thus most limiting to growth. Interestingly though, many of our better deciduous forest collections have come from straight slopes rather than from the more extreme convex ridgetops.

Dendroclimatological studies seek robust correlations throughout the range of factors to which growth is correlated. It would be valuable, for example, for both drought years and wet years to correlate equally well with growth, thereby not limiting reconstructions to one extreme or the other. However, some of the most extreme habitats may often yield data that correlate with only part of the range of the environmental factor under consideration. For example, a convex slope site might yield tree rings that do not correlate well with years of low amounts of precipitation. *Normal* conditions of an extreme convex slope site may be quite dry. Thus, water availability and growth at such a site during an extremely dry year (drought) may differ little from a more normal year. In that case, an extreme convex slope may actually support trees with rings that correlate well with greater than normal amounts of precipitation but poorly with droughts. Conversely, concave slopes may generally correlate poorly with the full range of precipitation, but may correlate nicely with droughts. Tree rings from straight slopes may not correlate with droughts quite as well as do those from concave slopes nor quite as well with greater amounts of precipitation as expected from trees on convex slopes. However, because they correlate reasonably well with both large amounts of precipitation and low amounts of precipitation, trees growing on straight slopes are often preferred for dendroclimatological studies.

Very powerful and sophisticated statistical methods have been used to describe correlations between growth and environmental factors. The intent is to more accurately and thoroughly describe a relationship where a correlation exists. On the other hand, if no correlation exists, no amount of fancy statistics and *data massaging* will create one. A very simple method to determine if a relationship exists is to simply create a scatter plot between growth (usually width index) and the factor in question (such as precipitation). For example, a scatter plot between tree-ring indices of continuously inundated baldcypress

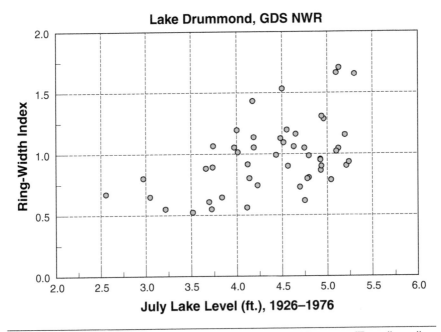

Figure 8.1   Scatter plot between tree-ring indices of baldcypress (*Taxodium dis-tichum* (L.) Rich) growing in Lake Drummond in the GDS and the mean lake level for July 1926–1976. See text section, The Lake Drummond Cypress.

(*Taxodium distichum* (L.) Rich.) growing in Lake Drummond in the Great Dismal Swamp (GDS) and the mean July lake level (Figure 8.1) suggests a positive correlation between tree rings and lake level.

Selecting Species

The Eastern Deciduous Forest is essentially an oak forest. Continual disturbance may tend to favor a ubiquitous species such as red maple (*Acer rubrum* L.). That notwithstanding, most trees old enough to permit a reasonable extension of the climatic record are still expected to be oaks. Taxonomically, oak species may be divided into a white oak group (*Lepidobalanus*) and a red oak group (*Erythrobalanus*). Generally, trees of the red oak group tend to be faster growing, thus resulting in wider rings with less ring-to-ring variation (smaller *MS* values) than are typical of the white oak group. White oak (*Quercus alba* L.) is by far the most common of the oak species and often is the species of choice in Eastern tree-ring work.

Some studies allow the freedom of choosing from among several species. In those cases, it might be best to extract a few sample cores from each species and simply compare them. Preference is usually given to older trees with reasonable ring-to-ring width variation that do not have obvious periods of disturbance. In some cases it might make sense to use several species in a multiple regression study, the rationale being that each species may respond sufficiently different enough that each contributes unique information to the whole.

Some studies involve the use of a specific species or a variety of species. In those cases the question may become, "Where can I find the best site for collecting this species?" This, in turn, often becomes, "Where can I find a stand of old trees of this species?"

Selecting Sites

Dendroclimatological studies attempt to construct collections that are representative or typical of the climatic region. A proposed study becomes a matter of finding stands of old trees from sites that are sufficiently limiting to yield good correlations between growth and climate. Interestingly, land that was left in forest (uncleared) from sites that were considered unsuitable for agriculture may be the very sites that are most limiting to tree growth and, conveniently, may also contain older trees.

Tree-ring collection sites for ecological studies often afford little choice in site selection. A study to examine growth responses to climatic factors of trees on north-facing versus south-facing slopes, for example, is restricted to using tree samples from those slopes regardless of how limiting they are to tree growth.

Selecting Trees

Gnarled and twisted trees growing on dry ridgetops are preferentially sought for study. There seem to be two or three reasons for this. Trees on ridgetops are expected to be old. At some sites, trees may actually have been left from the original, virgin forest. The hope is that they were not left because they were not sound, but rather because they were too twisted and bent to make good lumber. Most often, unfortunately, they were left because they were not sound. We have cored mammoth baldcypress in the GDS that were no doubt left from the original forest. Unfortunately, they were just shells. Some of these old giants were more than 2 m in diameter at sample height, but we were lucky to get a sound core greater than 10 cm in length.

On the other hand, a gnarled and twisted tree at a site that is not environmentally extreme may be nothing but trouble unless one knows the reasons for the tree being misshapen. If given the option, it is usually best to select trees that are not misshapen or leaning and to select trees with reasonably concentric crowns. A crown that is distorted because of excessive crowding results in an increased risk of missing rings if sampled on the suppressed side of the tree. If a tree has opposing crown radii that are excessively suppressed, then it may be best to skip that tree.

Selecting Sides of the Tree

There are those who will always collect by compass direction, such as on the north and south sides of the trunk or on the east and west sides. On the other hand, it may be far more important to sample beneath the least suppressed sides of the crown. Crowns are rarely perfectly symmetrical, but one can usually find nearly opposing sides of a trunk that are beneath crown segments that are not excessively suppressed from crowding. In broad generalities, widths of rings below the suppressed side of a tree crown tend to be more suppressed than rings below less suppressed sides of the crown. That is, chances are greater of encountering tiny rings with little ring-to-ring width variation, as well as missing rings.

An exception to collecting samples from the least suppressed sides is when a tree is growing on a steep slope. Typically, the downslope side of the crown will be more exposed than the upslope side. Consequently, slope trees normally lean down slope as a phototropic response. Leaning typically results in the formation of tension or compression wood that may be difficult to account for in data analysis. These may impart an error to ring-width data gathered for climatic information. The best way to avoid this is to sample along the diameter that is parallel to the contour lines of a topographic map; that is, along the slope. Selecting slope trees then becomes partly a matter of finding trees in which the crowns are not particularly suppressed on the sides along the slope.

THE LAKE DRUMMOND CYPRESS

Exploratory Research

Work in new areas often involves a certain amount of exploratory sampling to determine if the tree rings of local trees appear to be climatically sensitive. Initially in studies in the GDS, we did not know if rings of any trees in the swamp could be correlated with anything. As things progressed, we would occasionally sample something just to see if it seemed to display enough ring-width variability to potentially be climatically sensitive. To our surprise, we found that the rings of a continuously inundated baldcypress tree growing in Lake Drummond were highly variable. The following discussion describes information obtained from trees that we never expected could provide any information at all. To introduce the setting, we will begin with some background information.

The Lake

Lake Drummond is a natural black-water lake in the center of the GDS. The swamp is on the coastal plain along the border of North Carolina and Virginia (Figure 8.2). The lake is nearly circular, about 4.0 km in diameter, has a peat bank around its edge, and is only about 2 m deep. At the time of European settlement, there were no open waterways either feeding or draining the lake. The lake came to be used as a source of drinking water for ocean-going vessels because the black water of the lake remained *fresh*. The water contained such high concentrations of tannic acids that it was largely unsuitable for bacterial activity. Lake level at the time of European settlement, at least during winter and spring months, probably would have been an extension of the water level in the surrounding peat. (Though derived from sphagnum and referred to herein as peat, the material is not a true peat and hence, is more properly referred to as Dismal Swamp muck. Further, because of the prevalence of sphagnum moss, some people have argued that the GDS should be called a bog.)

George Washington visited the swamp in 1763. Subsequently, he formed a company that was commonly referred to simply as the Dismal Swamp Company. Work was soon begun to construct a hand-dug ditch extending from the escarpment that formed the western boundary of the swamp to Lake Drummond, about 8 km to the east in the center of the swamp. Perhaps the earliest use of the ditch, known now as Washington Ditch, was to float out cypress logs cut from the great cypress trees surrounding the lake. The Dismal Swamp Company hoped that Washington ditch would also serve to drain the swamp. Even though the swamp actually does slope a bit toward the east, the slope only averages about 20 cm/km. The Company did not apparently anticipate that the swamp was so flat.

The Dismal Swamp Canal Company, formed in 1784, started in 1793 to dig a north-south canal through the eastern part of the swamp. The intent was to connect the Chesapeake Bay in Virginia with the Albemarle Sound in North Carolina. While the Dismal Swamp Company was formed to tame the swamp and exploit its resources, the Dismal Swamp Canal Company was formed to shorten the distance needed to move supplies and goods along the coast. It took about 12 years of manual labor to build the 35 km canal. One result of the canal was to intercept the eastward movement of water across the swamp. Consequently, portions of the original swampland east of the Dismal Swamp Canal became dry enough to be suitable for agriculture. Though essentially flat, the central portion of the canal was elevated relative to the north and south ends, so that the central part of the canal was described as being little more than a muddy mess during large portions of the year. This was solved by installing locks and supplying water from Lake Drummond via a ditch, named the Feeder Ditch (Figure 8.2). The Feeder Ditch was completed in 1812.

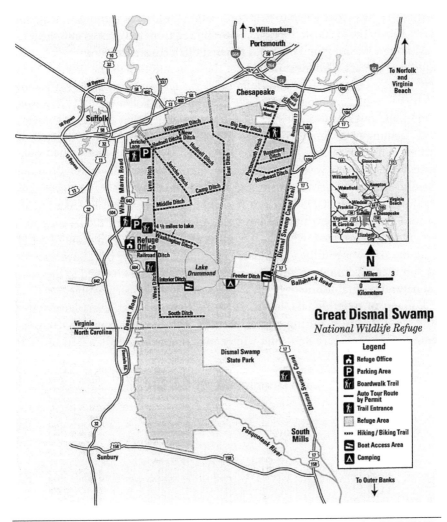

Figure 8.2   Map of the GDS National Wildlife Refuge (NWR) from the GDS NWR bulletin.

Additional ditch construction, in association with increasing timbering operations in the swamp, resulted in significant hydrologic changes to the swamp in general and to Lake Drummond in particular. Except for the very slight west to east gradient already mentioned, a natural flow gradient did not exist within the ditches or between the ditches and the lake. We have observed ditch water flowing slowly in opposite directions away from the location of a local summer shower. Establishment of many kilometers of ditches had the effect of increasing the surface area of the lake by the surface area of all the ditches. When water was released via the Feeder Ditch to help control the lake level or withdrawn from the lake to supply the locks on the canal, water flowed from the ditches into the lake.

Although Washington Ditch was the first ditch constructed, others followed in just a few years. The most notable of these, in addition to the Feeder Ditch, were Jericho Ditch (between Suffolk and Lake Drummond) and Portsmouth Ditch (between Portsmouth and Lake Drummond). Following WWII, Union Camp built an extensive network of ditches by which they hoped to enhance a tree farming operation through the use of water level regulation. After these ditches were constructed, but before the water regulation system was implemented, the property was obtained by the Nature Conservancy and then transferred to the Federal government and became the GDS NWR. At the time of establishment of the GDS NWR, a few old cypress remained in Lake Drummond. These old relics of the original ring of cypress around the lake are now continuously inundated, whereas it is most probable that at the time of settlement, the ring was routinely inundated in winter and spring but above water in late summer. Hydrologic conditions since settlement have changed to the point that natural re-establishment of the cypress ring is no longer possible.

A graph of monthly lake levels, 1926–1976, shows that levels have stayed fairly close to 1.6 m during winter months with marked summer decreases being common (Figure 8.3). The winter maximums are regulated in order to

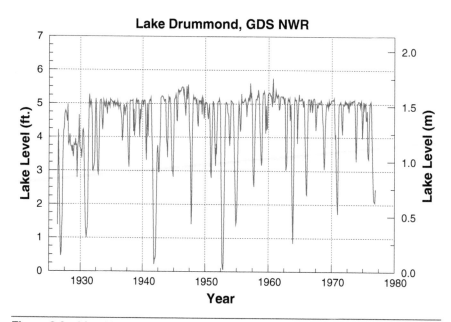

Figure 8.3 Monthly lake level of Lake Drummond, 1926–1976, averaged from daily data (U.S. Army Corp of Engineers).

control structures on the Feeder ditch. There are two short periods in which winter maximums got a bit higher (c. 1945–1947 and c. 1956–1961). Even these excursions only exceed the more typical years by 6 inches or less (≤ 15 cm). More outstanding though, is a declining trend in winter levels from 1928–1930. This stands out because the annual winter maximums as well as the summer minimums decline.

The Tree Rings

In the mid-1970s, I (RLP), with the assistance of Lee Applegate, took increment core samples from most of the cypress remaining in the lake. Obviously, these few trees had been left because they were not sound. At the time that they were sampled, most of the trees were simply hollow shells; hence, the tree rings contained on each core represented only a small fraction of the rings that would have been included on a full, intact radius. The amount of sound wood and the number of rings contained thereon varied considerably from core to core. Increment cores were surfaced and crossdated, then ring widths were measured to the nearest 0.01 mm. The number of rings and the inside dates of the core samples with the most included rings are shown in Table 8.1.

Table 8.1 Basic statistics of increment cores collected from cypress remaining in Lake Drummond. [Only those dating back to 1926 are included. Outside date of all samples is 1976. * = sample included in 8-core subset dating back to 1902. *ID* = inside date. *MS* = mean sensitivity.]

Sample	Rings	*ID*	Width (cm)	*MS*	Serial *R*
L031	63	1914	0.142 ± 0.109	0.787	0.494
L032	68	1909	0.132 ± 0.076	0.545	0.392
L041	53	1924	0.120 ± 0.078	0.417	0.408
L042	58	1919	0.127 ± 0.097	0.370	0.537
L050	57	1920	0.050 ± 0.023	0.303	0.144
L061	69	1908	0.136 ± 0.189	0.392	0.551
L062*	123	1854	0.133 ± 0.106	0.484	0.487
L071*	76	1901	0.125 ± 0.134	0.386	0.602
L072*	75	1902	0.208 ± 0.220	0.359	0.044
L081*	103	1874	0.114 ± 0.062	0.366	0.530
L082	66	1911	0.071 ± 0.058	0.484	0.433

Continued

Sample	Rings	ID	Width (cm)	MS	Serial R
L090*	138	1839	0.081 ± 0.064	0.462	0.341
L100*	75	1902	0.118 ± 0.076	0.642	0.011
L200*	80	1897	0.061 ± 0.037	0.457	0.420
L210	63	1914	0.168 ± 0.076	0.293	0.203
L220	61	1916	0.051 ± 0.028	0.664	-0.109
L231	64	1913	0.097 ± 0.089	0.392	0.420
L232*	135	1842	0.064 ± 0.067	0.461	0.411
L260	67	1910	0.205 ± 0.174	0.571	0.190

Trends in growth can best be described in terms of BAI. Some growth declines described in the previous chapter that were detected when graphed as BAI were not at all apparent when graphed simply as ring width. The cypress of concern here are a special case. A cross section of any given tree would have shown the outside shape to be a knobby circle caused by several lobes. The lobes, in turn, prevented the estimation of a meaningful radius length for use in calculations of BAI.

Ring-width trends were identified by smoothing the ring-width series with a cubic spline (Figure 8.4). Although only two cores are shown, most other cores had similar patterns of growth. The trending curves show a distinct decline until about 1900, an increasing trend until the mid-1940s, followed by another declining trend. It might be expected that periods of declining growth took place as a result of continuous inundation and that, conversely, the 1900–1950 period of increasing growth resulted from less than continuous inundation during the growth season. However, the lake level data do not bear this out. Observations of lake levels during midsummer of 1976 (just above 2 feet, see 1976 in Figure 8.3) showed that lake levels were not below the surface on which the cypress were growing. The number of years in which the lake level got below 2 feet (0.6 m), which may or may not have been below the level of the cypress, does not seem to differ during the 1926–1950 period than after 1950. The Feeder Ditch was deepened and widened around 1900, and Union Camp activity to repair existing ditches and construct new ones that interconnected with the lake began in earnest after the end of WWII. Although these two ditching activities may have occurred at about the same time as the 1900 and 1950 growth trend inflections, any kind of cause and effect relationship is not immediately apparent. For the time being, then, the causes for the growth shifts around 1900 and around 1950 will have to remain in an "unknown" category.

Before collecting the samples, we expected that the increment cores would show extremely tiny rings with essentially no ring-to-ring variation in width. Contrary to expectations, the MS measurements (Table 8.1) indicate considerable width variation, suggesting at least the possibility of correlation with

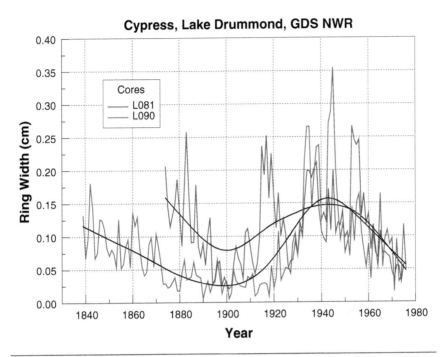

Figure 8.4   Ring-width series of two core samples, L081 and L090. Ring-width trends (black curves) were determined with a 60-year cubic spline.

environmental factors. Generally, factors are not expected to be correlated with growth unless they are limiting to growth. To begin with, lake level was excluded from consideration because if the root systems were continuously inundated, then lake level (an indication of depth of inundation) was expected to be inconsequential. Regardless of whether the lake level was high or low, the root systems were totally inundated. The tree couldn't care less, teleologically speaking, whether its roots were covered by 20 cm of water or 2 m of water. Similarly, precipitation was considered unimportant to trees under continuous inundation. Temperature, however, seemed like a plausible candidate, but correlations based on numerous time period combinations of temperature yielded nothing. Various precipitation time periods and a number of combinations of temperature and precipitation again did not yield meaningful results. Interestingly, a significant correlation between ring-width indices and the July lake level was produced. Why? There is no direct, obvious cause-and-effect relationship between tree growth and lake level. Tree growth was inferred to be causally related to something else that, in turn, was also causally related to the lake level.

Anyone who has been on black water on a still, cloudless day in mid-summer is quite aware that the air above the water can become exceedingly hot. The heat load on the black surface must result in considerable evaporative loss, thereby drawing down the water level in the lake. At the same time, the heat load on the cypress foliage could easily be great enough to curtail most physiological processes associated with growth, thereby resulting in reduced growth. Thus, it seems reasonable that a growth season with a greater than average amount of cloudless days might create enough of a heat load on the black-water surface that cypress growth is reduced and, simultaneously, lake level is reduced. This suggests that though lake level is not causally related to growth of the inundated cypress, lake level and tree growth are correlated because they are both causally related to the same factor—namely, excessive heat generated by solar radiation absorbed by a black-water surface.

Reconstructed Lake Level

The ring-width trends (Figure 8.4) were removed from the ring-width series by dividing ring widths by the corresponding values from smoothed ring-width curves, thereby producing tree-ring indices (Figure 8.5). The tree-ring indices of individual samples were used as predictors of the monthly lake level,

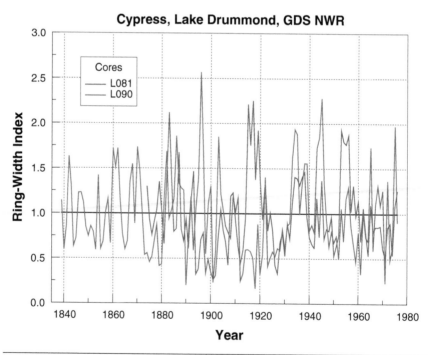

Figure 8.5   Tree-ring indices for the two increment core samples (L081 and L090) illustrated in Figure 8.4. MSL081 = 0.163 and MSL090 = 0.216.

1927–1976, in stepwise multiple regressions. A reasonably strong first order serial correlation was evident in the tree-ring indices (Table 8.1). That is, in addition to environmental conditions in any given year, growth in that year is influenced in part by how much growth took place the previous year.

Therefore, the previous year's index was used as well as that of the current year to predict current-year lake level. This meant that in the 19-core subset of cores dating back to 1926 (Table 8.1), a total of 38 variables were available for selection in the stepwise process. Lake-level data were available back to 1926, and the tree-ring samples contained rings only through the 1976 growing season. Because previous year indices were used as predictors along with current-year correlations for 1927–1976, R^2 values were adjusted for degrees of freedom lost according to the equation:

$$R^2 = 1 - \frac{n-1}{n-m-1}\left(1-R^2\right)$$

where:

    n = number of years, and
    m = number of variables selected.

Percent variance explained ($R'^2 \times 100$) ranged from 11% (Dec.) to 66% (Aug.); see Table 8.2. Interestingly, reconstructed lake levels (blue in Figure 8.6) were generally greater than actual lake levels prior to 1950 and less than actual levels after 1950. The relationship between lake levels and tree rings seemingly was not the same after 1950 than it was before 1950. Ring widths showed an increasing trend from about 1900 to the mid-1940s, followed by a sharp decline thereafter (Figure 8.4). This same pattern was observed among nearly all samples. Some factor seems to have affected tree growth differently between 1900 and the mid-1940s than between the mid-1940s and 1976. Thus, the two time periods were subsequently calibrated separately.

Calibrating tree rings with lake levels during the early period of record (1927–1951) and then using that relationship to reconstruct lake levels during the late period of record (1951–1976), Figure 8.7 shows a good match during the calibration period but essentially no match at all during the reconstructed period. Reversing this, tree rings of the late period of record were calibrated with lake levels and used to reconstruct levels during the early period of record (Figure 8.8). The two intervals of tree growth are sufficiently different that neither one can be used to estimate lake levels of the other. For illustrative purposes, only graphs of April lake levels are shown (Figures 8.7 and 8.8), but they may be considered as representative of the months showing the greatest correlations with lake levels, March, April, May, July, and August (Table 8.2).

Table 8.2   Number of variables selected and variance accounted for in regression by month and time period. [Previous year variables in parentheses.]

	1927–1976		1927–1951		1951–1976	
	Variables	**Percent**	**Variables**	**Percent**	**Variables**	**Percent**
January	4(3)	31.3	15(10)	96.7	2(0)	23.1
February	11(5)	62.2	9(2)	89.1	3(2)	33.2
March	6(1)	49.0	7(2)	82.8	7(4)	71.4
April	4(1)	54.0	10(2)	92.3	10(1)	92.6
May	3(0)	35.5	7(2)	83.8	4(2)	58.2
June	7(3)	42.2	2(0)	35.7	7(3)	74.5
July	7(3)	61.7	8(4)	76.2	8(3)	80.5
August	7(2)	65.5	7(4)	84.1	6(2)	67.2
September	6(2)	43.0	5(3)	65.6	5(0)	62.9
October	3(1)	22.4	6(4)	57.2	3(0)	37.5
November	9(2)	38.1	4(3)	57.2	2(1)	25.9
December	2(2)	11.0	4(0)	56.2	2(1)	17.0

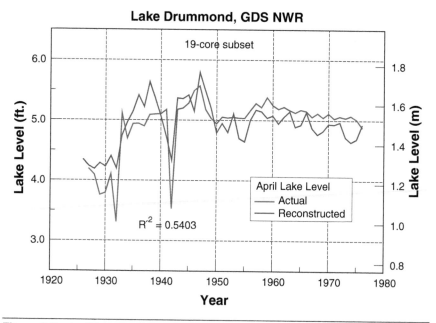

Figure 8.6   April lake-level data (red) and reconstruction of lake level from tree rings (blue). Note that the tree rings generally overestimate the lake level before 1950 and underestimate the lake level after 1950.

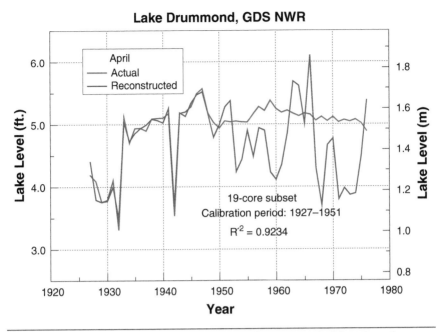

Figure 8.7   Actual lake level (red) and lake level reconstructed from tree rings (blue). Tree rings were calibrated with lake level for the early period of 1927–1951. The calibration-period regression was then used to reconstruct the lake level of the late period, 1951–1976. The relationship established for the early period obviously does not hold for the late period.

Calibration reconstructions (early period of Figure 8.7 and late period of Figure 8.8) appear almost too good to be real. This could be due to the stepwise process having selected an excessive number of variables, resulting in what is often referred to as over-fitting of the regression equation. It is possible that this may have been a result of using a prior year ring-width index that did not adequately compensate for the large degree of serial correlation indicated in Table 8.1. In these particular examples, the reconstructions beyond the calibration period have essentially no resemblance to the original lake-level data, and so are of no value. If precise reconstructions are needed, one way to more realistically account for the serial correlation would be the use of time series modeling such as autoregressive integrated moving average (or ARIMA) modeling.

These results serve to illustrate the importance of considering tree-growth trends when attempting to reconstruct an environmental (climatic) factor. Except for special cases such as the cypress under discussion, BAI may be more helpful in identifying a growth trend than ring width. If the growth trend during

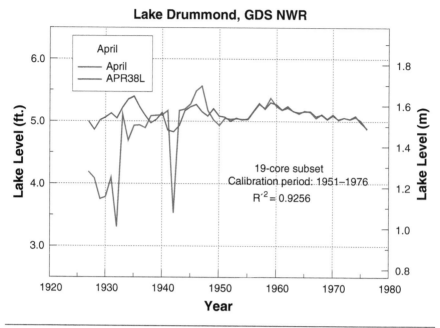

Figure 8.8   Actual lake level (red) and lake level reconstructed from tree rings (blue). Tree rings were calibrated with lake level for the late period of 1951–1976. The calibration period regression was then used to reconstruct lake level of the early period, 1927–1951. The relationship established for the late period does not hold for the early period.

the reconstructed time period differs from that of the calibration period, there is a good chance that the reconstruction will not be reliable.

The 1926 to mid-1940s interval was during an increasing growth trend that began around 1900 (Figure 8.4). This implies that the relationship between tree rings and lake levels remained consistent throughout this period, and that it might be possible to extend the lake level record backward in time before 1927. Because growth trends were different before than after 1900 (Figure 8.4), it would have been pointless to attempt to reconstruct lake levels earlier than 1900. Eight core samples dated back to 1902 (Table 8.1), which permitted reconstructions back to 1903. A new set of calibrations based on the 8-core subset (Table 8.3) was calculated for the 1927–1951 time period. Again, the ring-width index of the previous year was also included, meaning that 16 variables were available for selection in the stepwise process.

Using April lake levels as an example (Figure 8.9), the 8-core regression does not reproduce the lake-level record as faithfully as does the 19-core regression (Figure 8.6). That notwithstanding, the 8-core calibration was used to

Table 8.3 Number of variables selected (out of 16) and percent variance accounted for ($R'2 \times 100$) in stepwise regression estimations of the monthly lake level from tree rings (8-core subset) for the early period of increasing ring widths (1927–1951). [Number of previous year variables in parentheses.]

	Variables	Percent
January	3(1)	40.0
February	3(1)	43.7
March	3(1)	53.6
April	3(1)	60.2
May	3(1)	46.2
June	1(0)	19.7
July	4(2)	44.7
August	3(2)	53.3
September	4(2)	47.0
October	3(1)	42.6
November	6(3)	70.7
December	5(3)	58.0

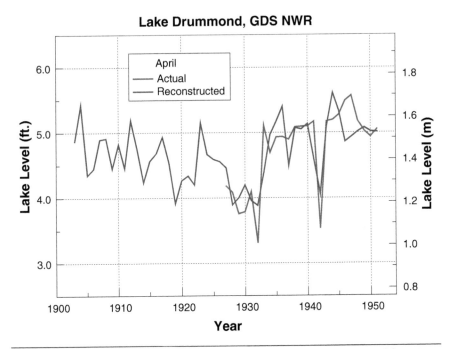

Figure 8.9 Actual lake level (red) and lake level that was reconstructed from tree rings (blue). Reconstruction is from the 1927–1951 calibration.

reconstruct lake levels for all months from the beginning of the period of record back to 1903. Even though the variance accounted for was reasonably strong (Table 8.3), reconstruction of some monthly values (namely for October, November, and December) were highly questionable and therefore deleted from the chronological sequence.

Reconstructed data suggest that lake levels before 1930 were higher and more variable than those after 1930. Summer lows do not seem as extreme before as after 1930 and there seems to be a decreasing trend in winter lake levels from about 1916–1925. Indeed, a declining trend in maximum winter levels begins well above 6 feet in 1916 and continues to decline to an average of about 4 feet by 1930 (Figure 8.10). Intuitively, it seems reasonable that water levels were manipulated due to release through the Feeder Ditch in the late 1910s and 1920s. Perhaps by around 1930, enough had been learned to enable combining anticipation with reaction, thereby permitting more predictable and stable maximum water levels. In keeping with that reasoning, perhaps the small increases noted

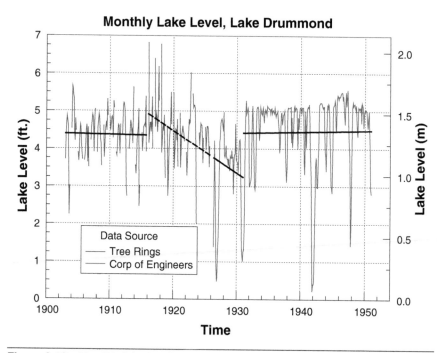

Figure 8.10 Monthly lake level, Lake Drummond, 1903–1950. Monthly data, 1926–1950 in red, from daily data supplied by the Corp of Engineers. Data for 1903–1925 (blue) reconstructed from tree rings of baldcypress (*Taxodium distichum* (L.) Rich) growing in the lake. Linear regressions of monthly data, shown in black, is presented to emphasize general trends.

previously in the late 1940s and late 1950s were brought about by the construction of additional ditches.

At this point, it is not known what caused lake levels to abruptly increase in 1916 and then decrease until 1930, before again increasing from 1931 to approximately the consistent levels maintained into the mid-1970s.

This exploratory investigation of the continuously inundated cypress in Lake Drummond is presented as an example of simple reconstructions of an environmental factor (lake level) from tree rings. The study yielded a few bits of information that might be useful in forming the basis for further investigation:

- Tree rings of continuously inundated cypress growing in Lake Drummond are discernable, measurable, and show good year-to-year variability.
- Ring widths of the lake cypress generally show a decreasing trend forward in time until about 1900, then an increasing trend to the late 1940s, followed by yet another decreasing trend.
- Tree-ring indices correlated positively with the monthly lake level for the years in common between lake level and tree-ring records (1926–1976). Better correlations were obtained when the record was divided into pre-1950 (increasing growth) and post-1950 (decreasing growth) periods.
- The positive correlation between baldcypress growth and lake level is not thought to be cause-and-effect; rather, it seems more likely that lowered water levels and decreased growth might both be related to the same variable, namely solar heat load on the black-water surface of the lake.
- An anomalous declining trend in the reconstructed lake level, 1916–1925, matched nicely with a similar trend in the Corp of Engineers' lake stage data, 1926–1930. Though certainly speculative, the trend is inferred to have resulted from lake level controls that were in effect during the period.
- Reconstructions suggest fewer constraints on maximum lake level prior to 1930, which in turn, suggests that bank formation by wave action may not have been well-defined until after 1930.
- Differences in the effects of environmental factors on tree growth before 1950 and after 1950 are suggested to be related to increased water input to the lake from a network of ditches constructed by Union Camp after WWII.
- Growth-trend projections suggest that had environmental conditions not changed after 1976, the cypress may not have survived past the year 2000. Thus, the fact that the lake cypress have survived implies that environmental conditions changed after 1976.

SELECTED REFERENCES

Great Dismal Swamp National Wildlife Refuge Brochure. U.S. Fish and Wildlife Service.

Cook, E. R. and L. A. Kairiukstis. (1990). *Methods of Dendrochronology*. Kluwer Academic Publishers, Dordrecht, Netherlands. p. 394.

Fritts, H.C. (1976). *Tree Rings and Climate*. Academic Press, New York, NY. p. 567.

———. (2001). *Tree Rings and Climate*. Blackburn Press, Caldwell, NJ. p. 567.

U.S. Army Corp of Engineers water level data, Feeder Ditch control structure, Great Dismal Swamp.

APPENDIX:
GRAPHS OF 47
WHITE OAK COLLECTIONS

People who assembled the ring-width collections are indicated in Table 7.3 of the text. The graphs are presented alphabetically by collection name. Each graph is the mean collection basal area increment (BAI) after 1850 calculated from the original ring-width collections. The number of cores represented by each graph is indicated, and the year by which all trees were in the canopy, generally 1900, is indicated. All cores extend the full length of the graph. The periods of change in slope (determined by regression) are years in which the second differences in slope are > 0.01 and are indicated graphically as a dashed line. Only changes in slope > 0.1 are indicated. Slopes of the periods between changes were determined by linear regression. The initial slope (graphed as black) of each collection is the first linear segment after all the trees are of canopy size (usually 1900). If a change in slope results in an increased slope, the new slope is graphed in blue. If the slope decreases > 0.1, the new slope is indicated in red. The same axis scales were used where practicable so that graph shapes are comparable among graphs. The Y-axis range for most graphs is 0–40, though a few may be 5–45 or 10–50.

- **1850**: Though many collections may have included years before 1850, they were only graphed back to 1850.
- **1900**: Target date for reaching canopy size. For most collections, only cores ≥ 10 cm by 1900 were included. For several collections the majority of the cores were from trees in the canopy well before 1900.
- **1950**: The approximate time of initiation of growth decline.

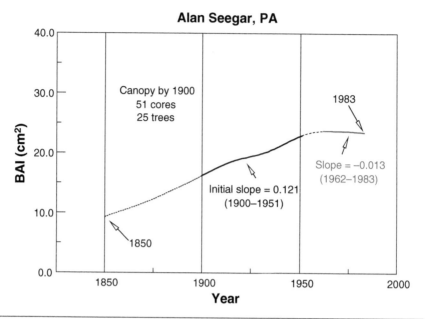

Figure A.1   Alan Seeger, Pennsylvania

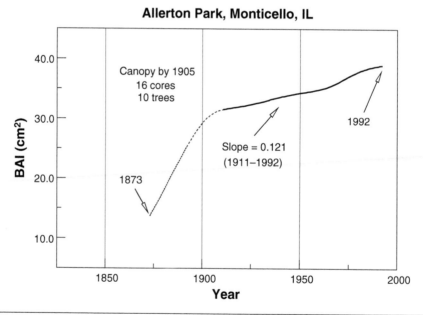

Figure A.2   Allerton Park, Monticello, Illinois

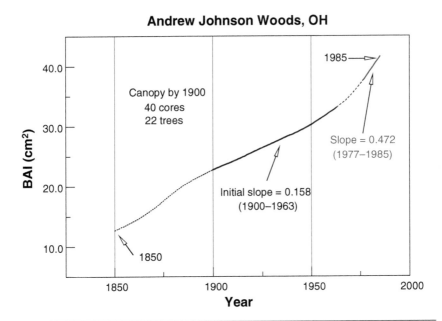

Figure A.3 Andrew Johnson Woods, Ohio

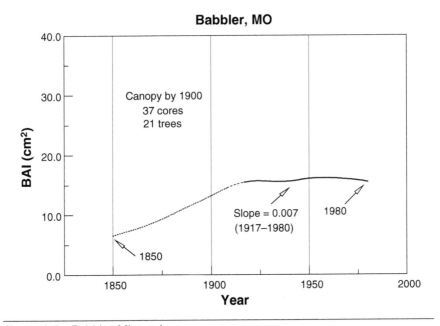

Figure A.4 Babbler, Missouri

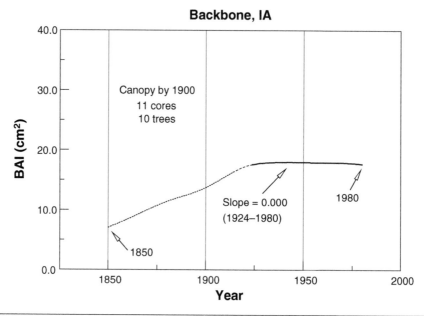

Figure A.5   Backbone, Iowa

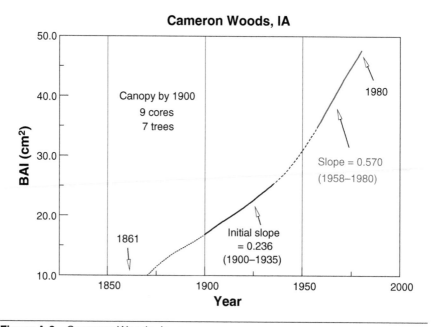

Figure A.6   Cameron Woods, Iowa

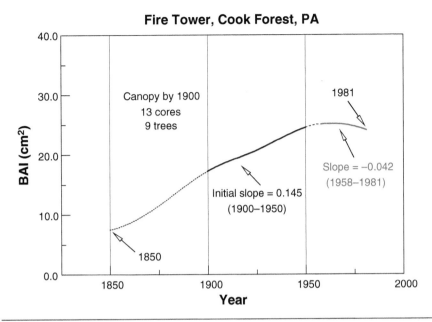

Figure A.7   Cook Forest Fire Tower, Pennsylvania

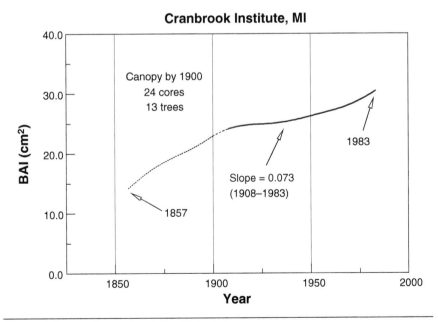

Figure A.8   Cranbrook Institute, Michigan

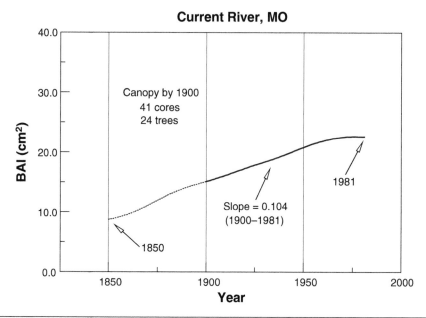

Figure A.9   Current River, Missouri

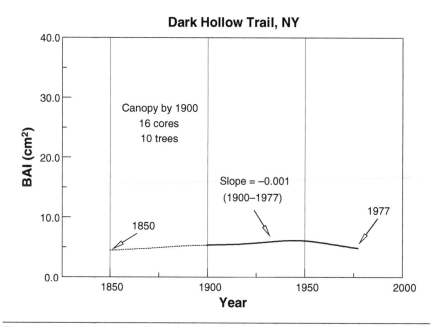

Figure A.10   Dark Hollow Trail, New York

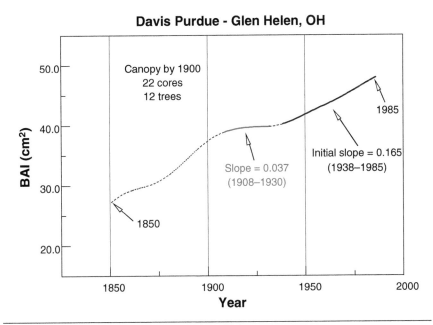

Figure A.11   Davis Purdue—Glen Helen, Ohio

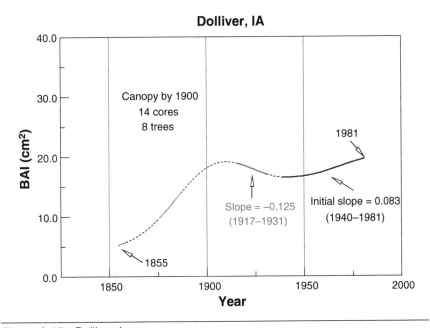

Figure A.12   Dolliver, Iowa

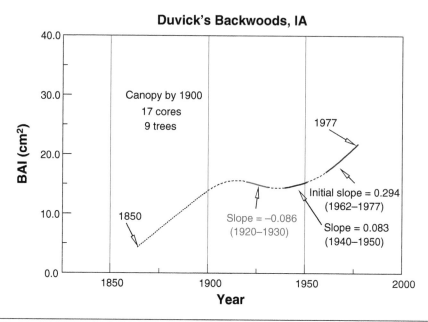

Figure A.13   Duvick's Backwoods, Iowa

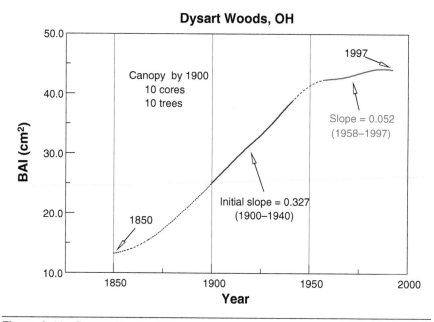

Figure A.14   Dysart Woods, Ohio

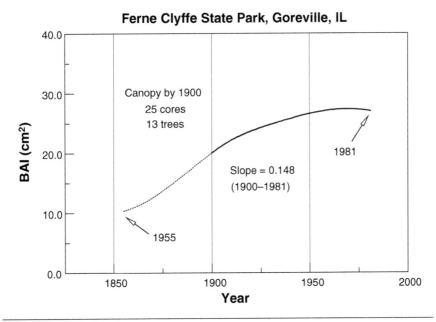

Figure A.15 Ferne Clyffe, Illinois

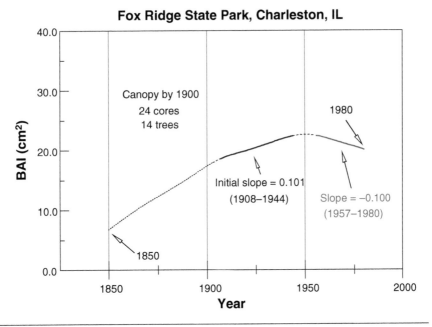

Figure A.16 Fox Ridge, Illinois

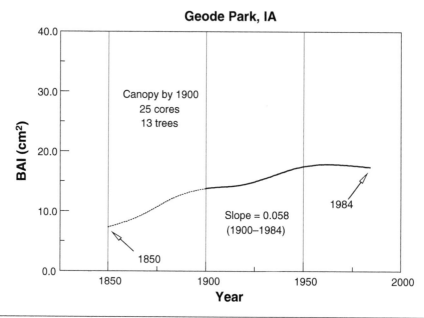

Figure A.17   Geode State Park, Iowa

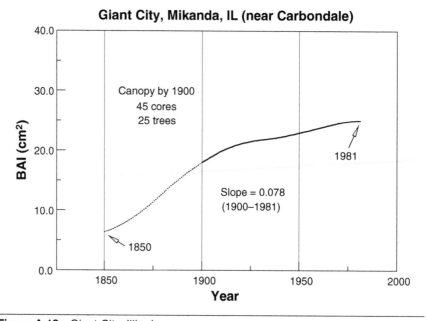

Figure A.18   Giant City, Illinois

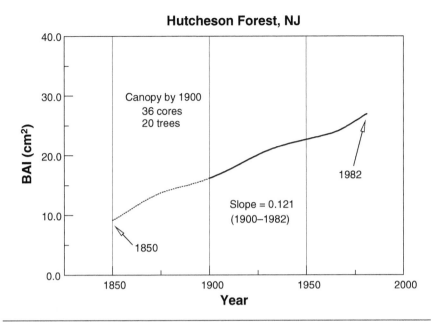

Figure A.19   Hutcheson Woods, New Jersey

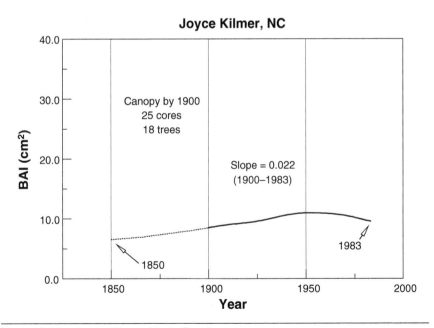

Figure A.20   Joyce Kilmer, North Carolina

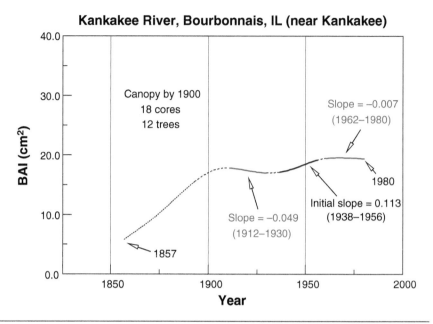

Figure A.21   Kankakee, Illinois

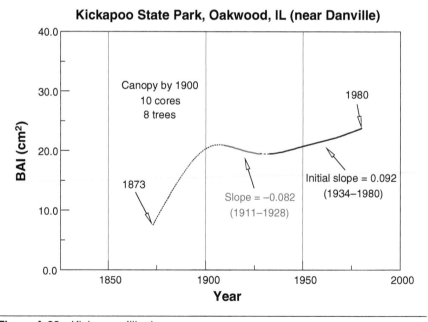

Figure A.22   Kickapoo, Illinois

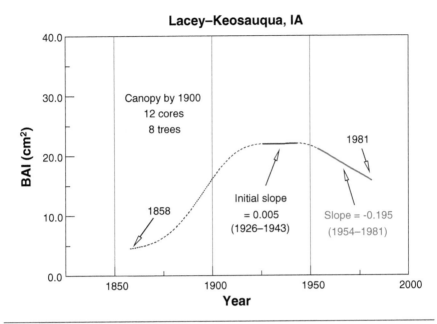

Figure A.23   Lacey–Keosauqua, Iowa

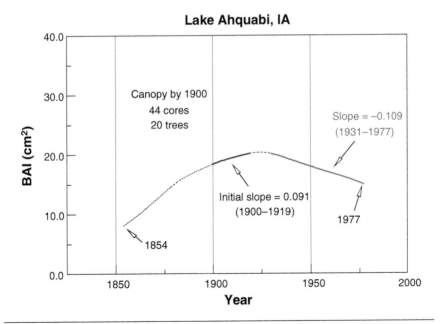

Figure A.24   Lake Ahquabi, Iowa

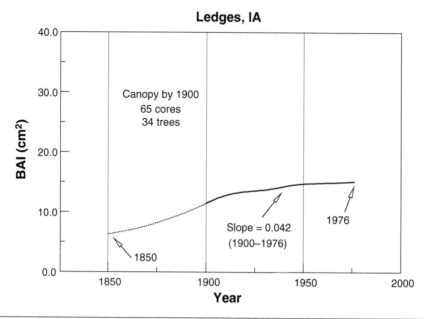

Figure A.25   Ledges, Iowa

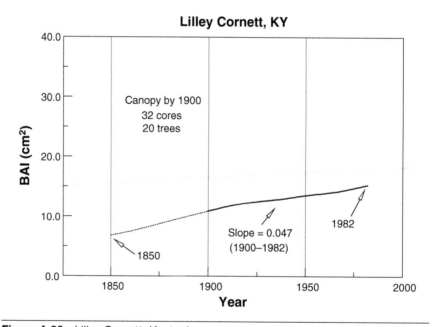

Figure A.26   Lilley Cornett, Kentucky

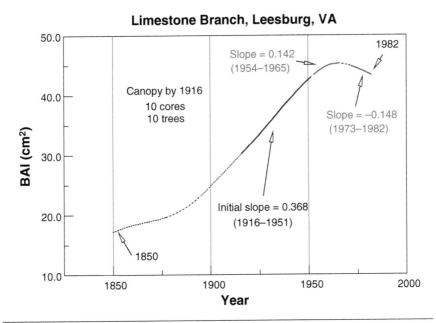

Figure A.27   Limestone Branch, Virginia

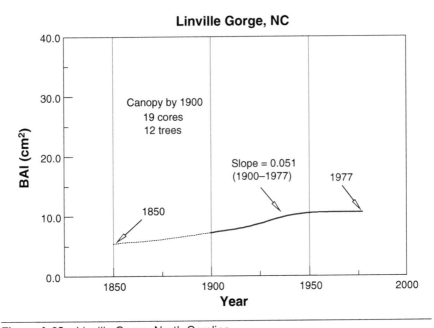

Figure A.28   Linville Gorge, North Carolina

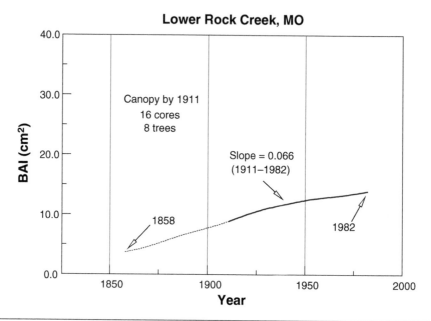

Figure A.29 Lower Rock Creek, Missouri

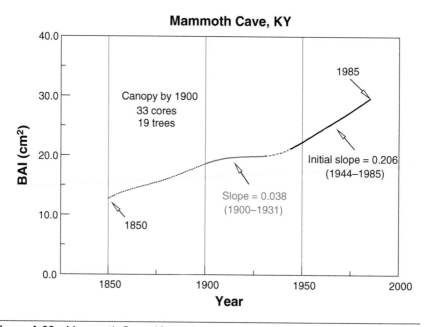

Figure A.30 Mammoth Cave, Kentucky

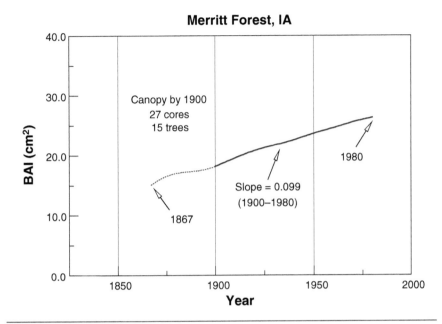

Figure A.31   Merritt Forest, Iowa

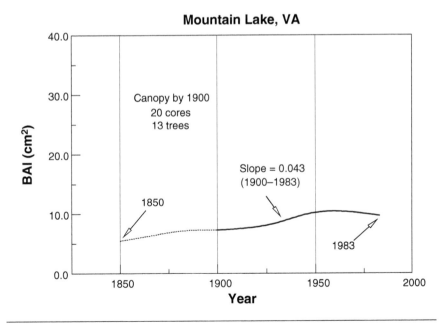

Figure A.32   Mountain Lake, Virginia

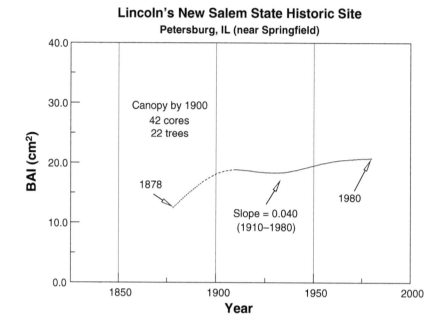

Figure A.33   New Salem, Illinois

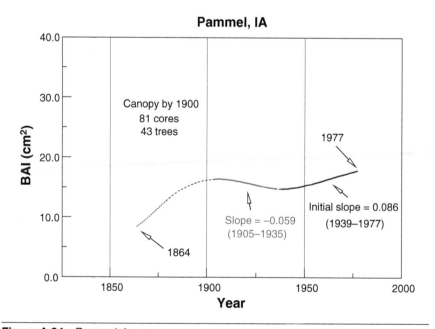

Figure A.34   Pammel, Iowa

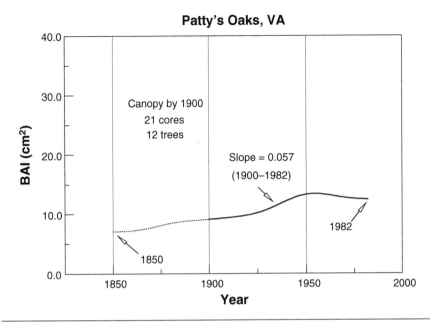

Figure A.35   Patty's Oaks, Virginia

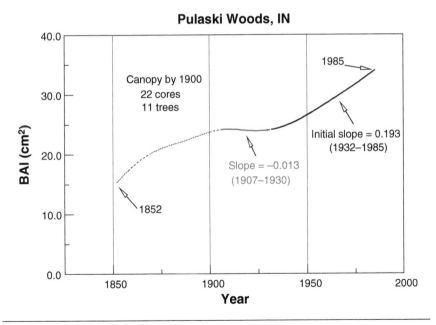

Figure A.36   Pulaski Woods, Indiana

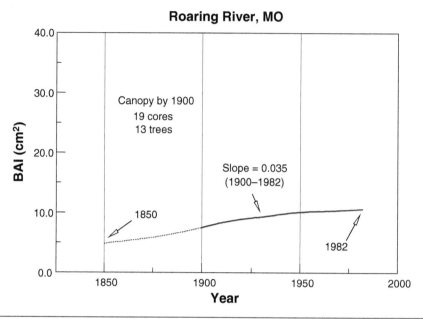

Figure A.37   Roaring River, Missouri

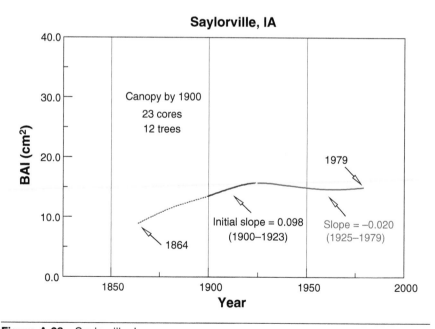

Figure A.38   Saylorville, Iowa

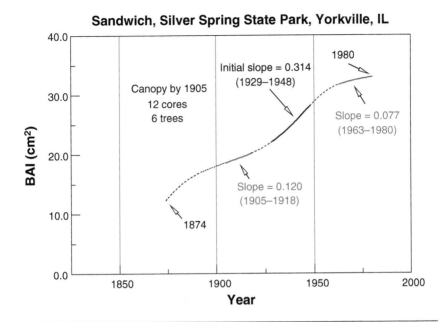

Figure A.39   Sandwich, Illinois

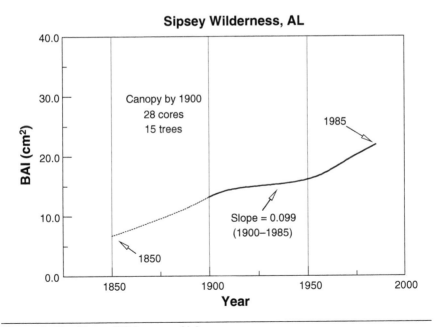

Figure A.40   Sipsey Wilderness, Alabama

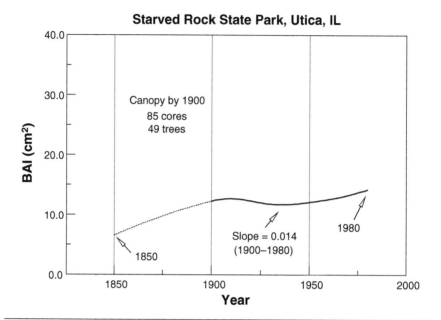

Figure A.41   Starved Rock, Illinois

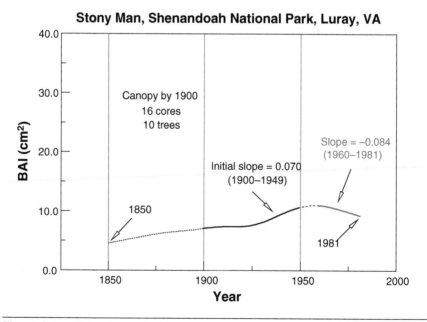

Figure A.42   Stony Man, Virginia

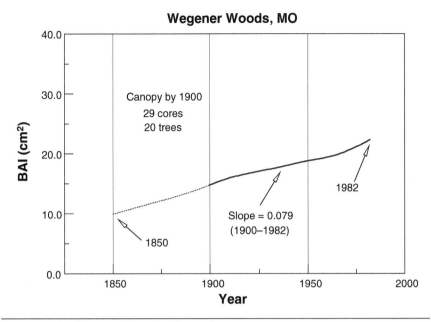

Figure A.43   Wegener Woods, Missouri

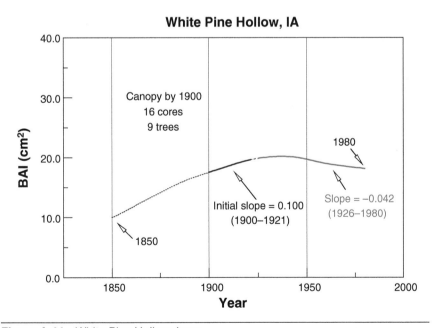

Figure A.44   White Pine Hollow, Iowa

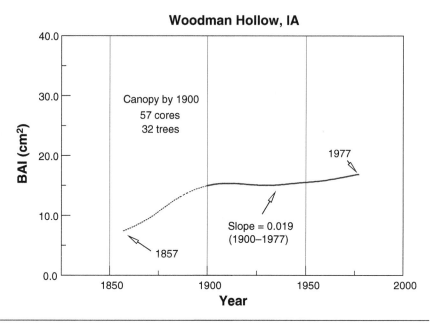

Figure A.45   Woodman Hollow, Iowa

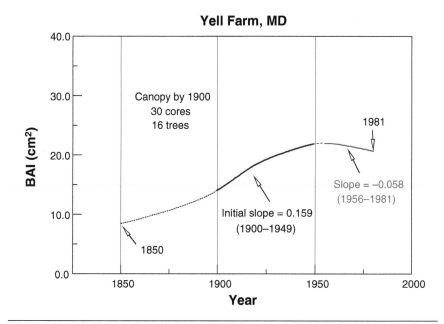

Figure A.46   Yell Farm, Maryland

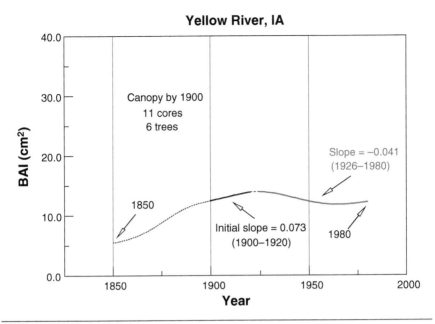

Figure A.47   Yellow River, Iowa

Table A.1   Slopes and changes. Regular type = under canopy (*radius* = <10 cm).
Bold = significant decrease (change > 0.1 slope). <u>Underline</u> = Initial slope (after
canopy closure). **_Bold italics_** = significant increase.

	Linear segment		Change period
1. Alan Seeger, PA	1850–1951	0.137	
	<u>1900–1951</u>	<u>0.121</u>	1952–1961
	1962–1983	**−0.013**	
2. Allerton Park, IL	1873–1892	0.532	1893–1910
	<u>1911–1992</u>	<u>0.121</u>	
3. Andrew Johnson Woods, OH	1850–1963	0.175	
	<u>1900–1963</u>	<u>0.158</u>	1964–1976
	_1977–1985_	**_0.472_**	
4. Babler State Park, MO	1850–1908	0.141	1909–1916
	<u>1917–1980</u>	<u>0.007</u>	
5. Backbone State Park, IA	1850–1914	0.137	1915–1923
	<u>1924–1980</u>	<u>0.000</u>	
6. Cameron Woods, IA	1861–1935	0.225	
	<u>1900–1935</u>	<u>0.236</u>	1936–1957
	_1958–1980_	**_0.570_**	

Continued

	Linear segment		Change period
7. Cook Forest (Fire Tower Road), PA	1850–1950	0.150	
	1900–1950	0.145	1951–1957
	1958–1981	**−0.042**	
8. Cranbrook Inst., MI	1857–1900	0.189	1901–1907
	1908–1983	0.073	
9. Current River Natural Area, MO	1850–1981	0.114	
	1900–1981	0.104	
10. Dark Hollow Trail, NY	1850–1977	0.009	
	1900–1977	−0.001	
11. Davis Purdue/Glen Helen, OH	1850–1860		
	1861–1867	0.092	1868–1880
	1881–1893	0.298	1894–1907
	1908–1930	**0.037**	1931–1937
	1938–1985	0.165	
12. Dolliver Memorial State Park, IA	1855–1859	0.126	1860–1876
	1877–1888	0.386	1889–1916
	1917–1931	**−0.125**	1932–1939
	1940–1981	0.083	
13. Duvick's Backwoods, IA	1864–1899	0.267	1900–1919
	1920–1930	**−0.086**	1931–1939
	1940–1950	0.083	1951–1961
	1962–1977	0.294	
14. Dysart Woods, OH	1850–1856	0.077	1857–1867
	1868–1940	0.322	
	1900–1940	0.327	1941–1957
	1958–1997	**0.052**	
15. Ferne Clyffe State Park, IL	1855–1981	0.140	
	1900–1981	0.148	
16. Fox Ridge State Park, IL	1850–1902	0.208	1903–1907
	1908–1944	0.101	1945–1956
	1957–1980	**−0.100**	
17. Geode State Park, IA	1850–1984	0.080	
	1900–1984	0.058	
18. Giant City State Park, IL	1850–1981	0.147	
	1900–1981	0.078	
19. Hutchenson Forest, NJ	1850–1982	0.126	
	1900–1982	0.121	

Continued

	Linear segment		Change period
20. Joyce Kilmer, NC	1850–1983	0.035	
	<u>1900–1983</u>	<u>0.022</u>	
21. Kankakee River State Park, IL	1857–1893	0.275	1894–1911
	1912–1930	**−0.049**	1931–1937
	<u>1938–1956</u>	<u>0.113</u>	1957–1961
	1962–1980	**−0.007**	
22. Kickapoo State Park, IL	1873–1881	0.591	1882–1910
	1911–1928	**−0.082**	1929–1933
	<u>1934–1980</u>	<u>0.092</u>	
23. Lacey–Keosauqua State Park, IA	1858–1864	0.139	1865–1889
	1890–1900	0.468	1901–1925
	<u>1926–1942</u>	<u>0.005</u>	1943–1953
	1954–1981	**−0.195**	
24. Lake Ahquabi State Park, IA	1854–1878	0.272	1879–1884
	1885–1919	0.118	
	<u>1900–1919</u>	<u>0.091</u>	1920–1930
	1931–1977	**−0.109**	
25. Ledges State Park, IA	1850–1976	0.079	
	<u>1900–1976</u>	<u>0.042</u>	
26. Lilley Cornett Tract, KY	1850–1982	0.063	
	<u>1900–1982</u>	<u>0.047</u>	
27. Limestone Branch, VA	1850–1877	0.084	1878–1895
	1896–1951	0.357	
	<u>1916–1951</u>	<u>0.368</u>	1952–1953
	1954–1965	**0.142**	1966–1972
	1973–1982	**−0.148**	
28. Linville Gorge, NC	1850–1977	0.050	
	<u>1900–1977</u>	<u>0.051</u>	
29. Lower Rock Creek, MO	1858–1982	0.088	
	<u>1911–1982</u>	<u>0.066</u>	
30. Mammoth Cave, KY	1850–1931	0.097	
	1900–1931	0.038	1932–1943
	<u>1944–1985</u>	<u>0.206</u>	
31. Merritt State Forest Preserve, IA	1867–1980	0.099	
	<u>1900–1980</u>	<u>0.099</u>	
32. Mt. Lake, VA	1850–1983	0.038	
	<u>1900–1983</u>	<u>0.043</u>	

Continued

	Linear segment		Change period
33. Lincoln's New Salem State Park, IL	1878–1893	0.286	1894–1909
	1910–1980	0.040	
34. Pammel State Park, IA	1864–1879	0.286	1880–1904
	1905–1935	**−0.059**	1936–1938
	1939–1977	0.086	
35. Patty's Oaks Blue Ridge Pkwy., VA	1850–1982	0.054	
	1900–1982	0.057	
36. Pulaski Woods, IN	1852–1858	0.331	1859–1869
	1870–1900	0.112	1901–1906
	1907–1930	**−0.013**	1931
	1932–1985	0.193	
37. Roaring River, MO	1850–1982	0.049	
	1900–1982	0.035	
38. Saylorville Dam, IA	1864–1923	0.114	
	1900–1923	0.098	1924
	1925–1979	**−0.020**	
39. Sandwich, IL	1874–1891		
	1892–1918	0.121	
	1905–1918	**0.120**	1919–1928
	1929–1948	0.314	1949–1962
	1963–1980	**0.077**	
40. Sipsey Wilderness, AL	1850–1985	0.099	
	1900–1985	0.099	
41. Starved Rock State Park, IL	1850–1980	0.041	
	1900–1980	0.014	
42. Stony Man, VA	1850–1949	0.049	
	1900–1949	0.070	1950–1959
	1960–1981	**−0.084**	
43. Wegener Woods, MO	1850–1982	0.088	
	1900–1982	0.079	
44. White Pine Hollow State Preserve, IA	1850–1921	0.138	
	1900–1921	0.100	1922–1925
	1926–1980	**−0.042**	
45. Woodman Hollow State Preserve, IA	1857–1886	0.203	1887–1891
	1892–1977	0.020	
	1900–1977	0.019	

Continued

	Linear segment		Change period
46. Yell Farm, MD	1850–1949	0.152	
	<u>1900–1949</u>	<u>0.159</u>	1950–1955
	1956–1981	**–0.058**	
47. Yellow River State Park, IA	1850–1920	0.135	
	<u>1900–1920</u>	<u>0.073</u>	1921–1925
	1926–1980	**–0.041**	

GLOSSARY

Absent ring: Missing ring.

Acropetal: Proceeding from the base toward the top. Contrast with basipetal.

Age trend: Time trend; growth trend; non-climatic component of ring-width or *BAI* series.

Angiosperm: Seed-bearing plants; broadleaf, hardwood trees.

Annual increment or *annual growth increment*: Three-dimensional sheath of secondary xylem (wood) added to stems and roots each year. The annual increment has the appearance of a ring (tree ring) when viewed in transverse (cross) section.

Annual ring: Tree ring.

Annual sensitivity (*AS*): Measure of ring-to-ring width variability; change in ring width from previous ring relative to mean of present and previous ring.

BAI: Basal area increment; tree-ring area.

BAI trend: Smoothed *BAI* data; that is, the non-climatic component of *BAI*; may be referred to as growth trend or time trend of *BAI*.

Basal area: Cross-sectional area at basal height, usually calculated from *dbh* measurement, but may also be obtained by summing *BAI*.

Basal area increment: Ring area increment calculated from ring width and radius length at basal height. *BAI* may also refer to an increment of *BA* obtained from periodic (usually not annual) measurements of *dbh*.

Basal height: Breast height.

Basipetal: Proceeding from top toward base. Contrast with acropetal.

Bole: Main stem, or trunk, of a tree.

Breast height: Chest height. Convenient height from which to take *dbh* measurements and increment core samples from trees, usually 1.3 or 1.4 m (4½ ft.).

Broadleaf: Bearing flat leaves, as opposed to needles. Broadleaf species are generally referred to as hardwoods. Most broadleaf species are deciduous.

Broadleaf evergreen: One of the few hardwood tree species, such as *Ilex opaca* Ait., that retain their leaves throughout the year.

Butt: The base of a tree trunk or main stem.

Calibration: A mathematical description (such as an equation derived from a multiple regression) of the relationship between tree-ring indices and the environmental factor(s) being investigated.

Cambial activity: The process by which cambial initials of the lateral meristem divide and form derivatives that differentiate into phloem cells outward from the cambium or xylem cells inward from the cambium.

Cambium or *cambial layer*: Lateral meristem; often several cells per radial file. A tissue in many higher plants, including trees, in which cells divide and form new tissues. The cambium surrounds stems and roots, and gives rise outward to the inner bark (phloem) and inward to the wood (xylem).

Canopy: The uppermost level of tree crowns in a forest. Tree crowns in the canopy are exposed to direct sunlight (solar radiation), as opposed to tree crowns in the understory which depend on transmitted sunlight and intermittent direct sunlight.

Chronology: Series of ring widths or indices placed in chronological order by year. Without modifiers, the term is often assumed to refer to a mean collection chronology.

Climatic component: Year-to-year variation in ring-width and *BAI* data, attributable to, or at least correlative with, climate; may be thought of as ring width or *BAI* with the non-climatic component (age or growth trend) having been removed. Ring-width indices are standardized estimates of the climatic component of ring widths.

Cookie: Informal forestry term for a very short segment of a stem or root; transverse section; cross section. A cookie is often used as a sample from which to examine tree rings on a transverse surface.

Conifer: Cone-bearing gymnosperm tree; softwood. Most coniferous trees, such as *Pinus* or *Tsuga*, retain their leaves (needles) throughout the year (*Taxodium* is an exception).

Core chronology: A chronology composed of data from a single increment core sample; a radius chronology measured from an increment core.

Core sampler: Usually refers to commercially available increment borers used to extract a core sample of wood; may also refer to specialty samplers such as might be used to obtain samples from building timbers at archaeological sites.

Crossdate: The process of assigning dates to a tree-ring series when matching or comparing tree-ring patterns with a dated series (dated chronology); matching with an undated series produces a floating chronology.

Date: Crossdate.

Dbh: Diameter at breast height. Often used as a measure of maximum tree stem (trunk) diameter exclusive of flaring at the base of the tree.

Deciduous: Annually loses leaves in the fall each year and over-winters in a leafless condition. Most deciduous species are hardwoods.

Deciduous conifer: Conifer that loses its leaves (needles) over winter, such as *Taxodium* or *Larix*. Contrast with evergreen hardwood.

Dendrochronology: The science of tree-ring dating to obtain exact year of each annual growth increment or tree ring; tree-ring science in general.

Dendroclimatology: A specialization of dendrochronology dealing with climate; the science of estimating, or reconstructing, climatic data from tree-ring records.

Dendroecology: Tree-ring studies concerned with ecological problems and applications. Precise definition may vary widely among authors.

Diffuse-porous: Wood type which contains pores, or vessels, relatively evenly distributed throughout each ring. Compare with non-porous and ring-porous.

Discontinuous ring: A tree ring which is not continuous around a given circumference. At interruptions where there is no ring, the ring is referred to as being missing or absent.

Earlywood: Wood tissue formed from the lateral meristem in the early part of the growing season. Cells in earlywood are larger and have thinner walls than in latewood. Earlywood may apply to all Temperate Region tree species; however, it is recommended that the term should be replaced by pore zone for ring-porous woods.

Evergreen: Retains green leaves during the dormant, over-wintering season, such as with *Pinus* or *Ilex*.

Evergreen hardwood: Broadleaf evergreen such as *Ilex opaca* Ait. or *Quercus virginiana* Mill. which retain green leaves over winter.

False ring: An extra ring within a true ring caused by the formation of a false ring boundary.

False ring boundary: Tissue appearing as a ring boundary before the true ring boundary such that continued growth after the false ring boundary appears as an additional ring; results in a false (multiple) ring.

Fibers, wood fibers, fiber elements: Xylem elements considered as non-water conducting; often difficult to distinguish from tracheids.

Floating chronology: Chronology for which dates of individual rings have not been established; chronology that has not been "tied down" by crossdating it with a dated chronology.

Forest level: Crown level; vegetative level; category of tree size based on relative height of the tree crown. Number of levels (≥ 2) varies with forest type and among authors:

Growth trend: Smoothed ring-width and smoothed *BAI*; an estimate of the non-climatic component of a tree-ring series.

Gymnosperm: Cone-bearing plants; conifers; needle-leafed softwoods. Most are evergreen.

Hardwood: Broadleaf angiosperm tree species; may refer to a whole forest (as the hardwood forest) or to individual trees or to the wood of such trees; usually deciduous.

Increment borer: Hand or power tool (specialized hollow wood bit) used for obtaining increment cores. Most commercially available increment borers are made from Swedish steel.

Increment core: Pencil-shaped wood sample obtained with an increment borer or increment hammer.

Index: Used without modifiers almost always refers to ring-width index; tree-ring index.

Inner bark: Phloem; food-conducting tissue of the tree.

Lateral meristem: Cambial tissue that envelops stems and roots and produces initials which enlarge and differentiate into xylem and phloem.

Latewood: Wood tissue formed in the latter part of the growth season; formerly, summerwood. Cells formed in late season are usually smaller and have thicker walls than cells formed in the early part of the season; hence, the last formed latewood, especially in conifers, may appear darker in color.

Marketable bole: The portion of the tree trunk that is suitable for sale. If the marketable bole is to be used for lumber, it is usually branch free and is cut into one or more logs.

Mean tree chronology: Mean of two or more core or radius chronologies for a given tree.

Mean collection chronology: A chronology obtained by averaging together by year the tree-ring data of all samples of a collection; often referred to simply as a chronology.

Mean sensitivity (*MS*): Mean of the absolute annual sensitivity values; standardized mean absolute change in ring width from ring to ring.

Meristem: Region of active cell division; may refer to either terminal or lateral meristem.

Missing ring: Absence of a ring; not represented by a tree ring at a particular point. A ring may be entirely missing or discontinuously (intermittently) missing around a given circumference.

Multiple ring: A tree ring including one or more false rings.

Non-climatic component: Ring-width trend or *BAI* trend; usually estimated by fitting a curve to ring-width data or *BAI* data; may be thought of as ring-width or *BAI* data from which the climatic component has been removed.

Non-climatic trend: Non-climatic component; growth trend; estimated as smoothed ring width or *BAI* trend.

Non-porous: Contains no pores or vessels. Contrast with diffuse-porous and ring-porous.

Outer bark: Outer layer of bark, arising from outer part of inner bark and from cork cambia within the outer bark.

Overstory: Uppermost tree crown layer; canopy layer. Contrast with understory.

Phloem: Food-conducting tissue of plants; in trees, the inner bark.

Photosynthates: Collectively, the various products of photosynthesis; food.

Pore zone: Zone of large pores (water tubes, vessels) at the beginning of rings in ring-porous woods. Evidence suggests that cambial initials are cut off from the cambium at the end of the previous season, remain as cambial initials over winter, and then enlarge and differentiate as large vessel elements in the current early spring. There is no counterpart to the pore zone in either diffuse-porous or non-porous woods.

Radius chronology: A chronology of tree-ring data from a single radius. The radius is usually identified on an increment core or a transverse section such as a cookie.

Raw basal area increment: Basal area increment that has been calculated directly from raw ring width, that is, ring width that has not yet been divided into climatic and non-climatic components. Thus, raw basal area increment (BAI_r) contains both climatic and non-climatic components.

Raw ring width: Original ring-width measurements.

Reconstruction: Estimation from tree rings of data of an environmental factor, usually used to extend climatic records for years prior to existing records.

Ring: Tree ring.

Ring boundary: The interface between consecutive rings, usually identified by a change in cell size or color resulting from a springtime resumption of growth after a winter dormant period.

Ring-area increment: Transverse area of a ring at any height; same as basal area increment except not limited to basal height.

Ring-porous: Wood in which there are pores, or vessels, of more than one size: small pores scattered throughout the ring as in diffuse-porous wood and large vessels found in a pore zone at the beginning of each ring. Compare with non-porous and diffuse-porous.

Ring width versus *ring-width*: No hyphenation when used as a subject, but hyphenation is used when the term is a modifier as in "ring-width indices are derived from measurements of ring widths."

Ring-width index: Standardized estimate of the climatic component of ring width; an index calculated as raw ring width relative to smooth ring-width trend.

Root collar: The transition zone between stem and root. Increment cores are often taken at breast height in order to be above any growth influence of the root collar.

Secondary xylem: Xylem produced from a meristem.

Semi-ring-porous: Wood type in which the large tracheary elements (vessels or pores) are not confined to a pore zone as in ring-porous woods; may appear intermediate between ring-porous and diffuse-porous woods. Examples include *Juglans* and some *Carya* (pecan hickories).

Sensitivity or *ring sensitivity*: See annual sensitivity (*AS*) and mean sensitivity (*MS*).

Sieve tube: Phloem food conducting tube formed by sieve tube cells arranged end to end.

Sieve tube cells: Individual cells which together form sieve tubes. A mature sieve tube cell loses its nucleus but, unlike a xylem vessel (water tube), remains alive.

Sieve plate: Perforated end walls of sieve tube cells.

Softwood: Gymnosperm wood or tree species.

Springwood: Old term for earlywood.

Standardized ring width: Ring-width index.

Sub-canopy: Tree crown level often considered as lower part of the canopy or overstory; the crowns are typically exposed to transmitted and discontinuous direct solar radiation.

Summerwood: Old term for latewood.

Terminal meristem: Region of active cell division at a stem or root tip that results in growth in length.

Tracheary elements: Water-conducting elements consisting of vessel elements and tracheids.

Tracheids: Water-conducting elements that are not open tubes or pores; tracheids may be quite difficult to distinguish from fibers.

Tree ring: Annual growth increment of xylem especially when viewed in transverse section. Though rings may be discerned in both xylem and phloem, without qualifiers, tree ring refers to a xylem ring.

Tree ring versus *tree-ring*: No hyphenation when used as a subject, but hyphenation is used when the term is a modifier as in "tree rings are used in a tree-ring study."

Tree-ring chronology: A chronology of tree-ring data.

Tree-ring data: Data derived from tree rings; usually refers to measured ring widths or data derived from measured ring widths such as ring-width indices, ring area, or basal area increments.

Tree-ring index: Ring-width index. Standardized estimate of the climatic component of a tree-ring series. Ring width with non-climatic component having been removed.

Tree-ring series: A chronological list of tree-ring widths or *BAI* values from a single radius.

Vessels and *vessel elements*: Water-conducting tubes or pores; tracheary elements exclusive of tracheids.

Understory: Vegetation beneath the forest canopy; includes small trees and shrubs as well as herbaceous species.

Wood: Secondary xylem in trees.

Wood fibers: See fibers.

Xylem: Water-conducting tissue in plants. In trees, the xylem is what is referred to as wood.

INDEX